"EX LIBRIS"

THIS BOOK

BELONGS TO: ____________________

____

PUBLISHED BY

'F.R.E.E.D.O.M.S.
MAGIC SQUARE STUDIO'

FOR
RESEARCH
EXPRESSING
ESSENTIAL
DATA
OF
MAGIC
SQUARES

mathemagics.

CORRESPONDENCE

JAIN
C/- 'MAGIC
SQUARE STUDIO'
777 LEFT BANK RD
MULLUMBIMBY CK
N.S.W. 2482
E: jainmagic42@
Hotmail.com

"All truth passes three stages: First, it is ridiculed. Second, it is violently opposed. Third, it is accepted as being self-evident."

Schopenhauer

The VEDIC MATHS CURRICULUM for the GLOBAL SCHOOL

JAIN

PART ONE
"DIGITAL SUMS"

2001

THE
VEDIC MATHEMATICS CURRICULUM
FOR THE
GLOBAL SCHOOL

PART 1 OF 4: the Galactic Mathematics of DIGITAL SUMS

ISBN: 978-1-925834-13-0

Snail Mail: 777 Left Bank rd, Mullumbimby Creek, 2482
Far Northern N.S.W. Australia.

Email: jain@jainmathemagics.com

Website: www.jainmathemagics.com

Introducing

THE VEDIC MATHEMATICS CURRICULUM FOR THE GLOBAL SCHOOL

BOOK ONE :

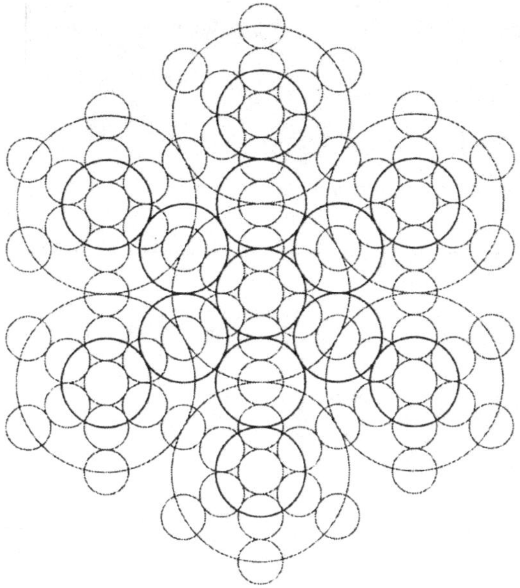

THE GALACTIC MATHS OF DIGITAL SUMS

Sub-Titled : " HOW TO TEACH YOUR CHILD THE ART OF VEDIC MATHEMATICS "

NEW RELEASE

ISBN for the ongoing Set of Titles is :

0-9594180-9-1

ISBN for this BOOK 1 is:

0-9594180-8-3

Over the next few years, the proposed ongoing Vedic Mathematics Curriculum workbooks for The Global School, will be published in this order :

- Book 1 - Digit Sums
- Book 2 - Multiplication
- Book 3 - Division
- Book 4 - Decimalisation of fractions
- Book 5 - etc

Contents

CONTENTS

MATHS WITHOUT TEARS

FOREWORD

by

JAIN

2001

an Ambassador for the Global School

"THE VEDIC MATHEMATICS CURRICULUM FOR THE GLOBAL SCHOOL' is intended as a first year text book for Teachers and Parents, to effectively teach Vedic Mathematics to children at Primary School or High School Level.

It is based on the fundamental principles of Vedic Mathematics which were reconstructed in the C20th by Sri Bharati Krishna Tirthaji (1884 - 1960).

These workbooks aim at providing a simple and clear mathematical training without the necessity of relying on the modern day calculator. If the next generation of children continue to use calculators for the simplest of computations, the global brain will enter a state of degenerative atrophy. Children are encouraged to build the 'Mental Muscle' by use of mental calculations.

Pattern Recognition is the key to understanding the natural and ancient laws expressed in the 16 Vedic Sutras that efficiently solve every known mathematical problem.

The methods presented here are fast and effective and rely on quick mental solutions that appear to be based upon "Intuition".

In effect, we are stimulating the shrunken Pineal Gland (at the very centre of the brain) or Third Eye, to re-swell to its natural size. Mystics tell us that the 3rd Eye was once the size of a large cherry, but now it is the size of a pea. This

FOREWORD

crystalline gland, to function properly, requires lots of pure drinking water (as the Brain is 90% water!) to help its growth transformation. It is the awakening of our natural clairvoyant vision that will enable our children to "see" the "Predicted Answer", by knowing "The Patterns Within".

When a mathematical problem can not be solved mentally, it can be reduced to a single line of working out. This Vedic ONE-LINE MENTAL METHOD is very helpful in computing, and checking results, as well as solving complex problems.

It is my earnest aim to bring this treasure trove of mathematical knowledge within the grasp of everyone who wishes to obtain it and

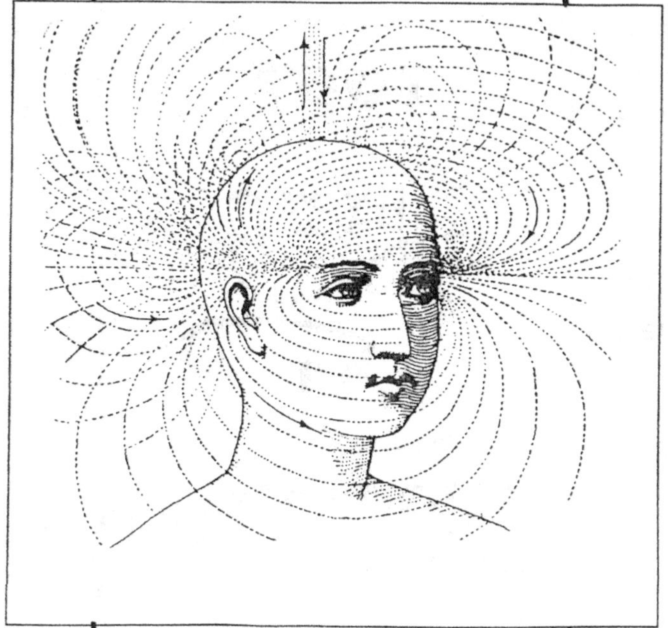

benefit by it.

One of the original meanings of the Greek word for Mathematics was:

"MATHEMATIKE" which meant "Learning". It implies the universality of number awareness in our Education, the Learning of Star Language hidden in number patterns and formulae. It was duly systematized and encoded into the Vedas, which as a compilation of texts, are the oldest storehouse of knowledge on Earth.

Unfortunately, on Earth today, there are many Teachers and Memory-Gurus charging people big money to learn memory skills, speed learning, even learning how to memorize whole telephone books of data! to cram more

FOREWORD

information in. This is dangerous. Its not about taking knowledge from "out there" + packing it in neurally to "In Here", which is a very male concept, from the Outside to the Inside :

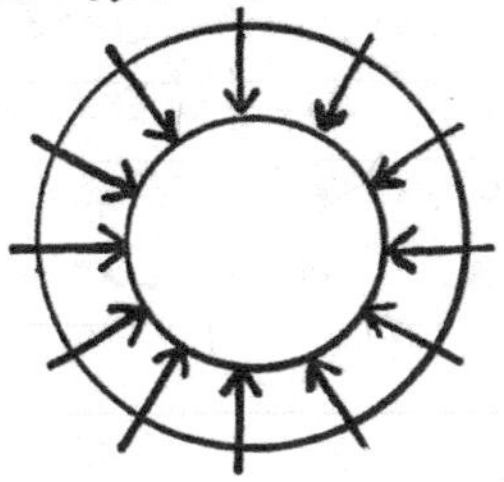

but rather, it is the opposite and feminine principle; its about

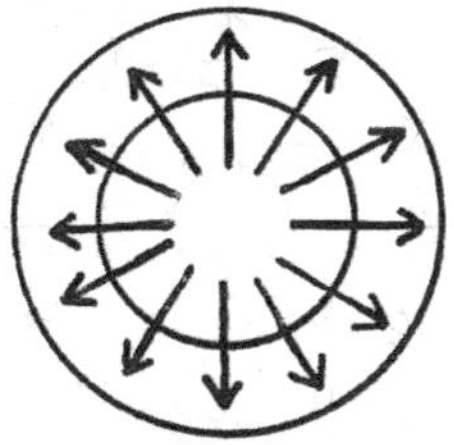

"unpacking", about moving from the Internal well-spring of "I KNOW" or Intuition to the External. Its really about

REMEMBERING that we are, and always have been, already tuned into the Vibrational Library of All Knowledge, known in the East as The AKASHIC (Spirit) Records. In fact, all memory of all Emotions, of all Lives, of all Deeds and Thoughts are encoded in your Cells, in The Living Library or the Repository of your Multi-Dimensional Self.

I believe that Vedic Maths is a Universal Language that undoubtedly triggers MEMory of All Knowledge. You already know All these Sutras. Take time to sit by a quiet creek and observe,

FOREWORD

in that stillness, the rays of the sunlight filtering down through the canopy, bouncing off the water's surface and reflecting back into your Eyes. Watch the Sacred Geometry of the interpenetrating rings of rippling waters expanding and telling you its Story, its Memory. Remember, "MEM" is the Hebrew word for "WATER" and "ORI" is the Hebrew word for "LIGHT". And as you contemplate upon this Image, a leaf spirals down poetically from the tree-tops and inspires you to Copy and Comprehend Nature. This is your true Teacher, not from books. The 16 Sutras are 16 Principles of how The Nature and The Mind work.

KNOW THYSELF.

Vedic Mathematics encourages:

- mental agility
- ambidexterity
- improves memory
- promotes efficiency

Self Confidence!

- no need for calculators
- appreciation and
- Self-Confidence

The Author
Jagadguru Śaṅkarācārya
Śrī Bhāratī Kṛṣṇa Tīrtha Mahārāja
(1884–1960)

Dedication

[The overall dimension of each page is in the Phi Ratio (1 : 1·618) or known here as The Golden Rectangle, which is the mathematics of DNA, Stars, Flowers and the Human Body. This Divine Proportion imbues a sense of aesthetic beauty to the Eye of the Beholder. Each large Golden Rectangle (170 x 275·06 mm) is subdivided into 2 smaller Golden Rectangles (170 x 105·06 mm) as shown as this border].

MAY THIS LIVING MATHS OF NATURE IMBUE OUR LIVES + OUR CHILDREN WITH UNITY CONSCIOUSNESS.

SRI BHARATI KRSNA TIRTHAJI - More than a Mathematical Genius

Sri Bharati Krsna Tirthaji (hereafter "Bharati Krsna") lived from 1884 to 1960. He is said to have reconstructed the ancient system of Vedic Mathematics from certain Sanskrit texts which other scholars had dismissed as nonsense. He tells us that the Vedic system which he rediscovered is based on sixteen Sutras which cover all branches of mathematics, pure and applied. The methods he showed and the simple Sutras on which it is based are extraordinarily simple and easy and the whole system possesses a unity not found in conventional mathematical methods. It can hardly be doubted that Bharati Krsna's remarkable discoveries in mathematics will in time change the teaching of and approach to mathematics world-wide: but this was not his main interest in life. His life was devoted to helping those individuals he could and also helping to bring about world peace and spiritual renewal. This short article aims to show something of his character and life and is drawn mainly from an introduction by Mrs Manjula Trivedi in the book "Vedic Mathematics" by Bharati Krsna. She looked after Bharati Krsna in the last years of his life and afterwards took charge of the Foundation he set up in Nagpur in 1953- the Sri Vishwapunarniman Sangha.

According to Manjula Trivedi, Bharati Krsna "named as Venkatraman in his early days, was an exceptionally brilliant student and invariably won the first place in all subjects in all the classes throughout his educational career. . . at the age of just twenty he passed M.A. Examination in further seven subjects simultaneously securing the highest honours in all, which is perhaps the all-time record of academic brilliance. His subjects included Sanskrit, Philosophy, English, Mathematics, History and Science". In 1908 he was made first Principal of the newly started National College at Rajmahendri, a post he held for three years.

Having a "burning desire for spiritual knowledge, practice and attainment" he then spent many years at the most advanced studies with the Shankaracharya at Sringeri in Mysore and was given the name Bharati Krsna Tirtha when he was initiated into the order of Samnyasa at Benares in 1919. He later, in 1925, became a Shankaracharya (the highest religious title in India).

He believed in the ancient Vedic tradition of all-round spiritual and cultural harmony, and his ambition for humanity was a world-wide cultural and spiritual renewal. People flocked to him in crowds and waited at his doors for hours. Granted an interview with him people felt that he immediately knew their need. Even when suffering from excessive strain he refused to take rest, continuing with his studies, talks, lectures and writings with unabated and youthlike vigour and enthusiasm.

In 1958 Bharati Krsna went on a tour to America, addressing audiences in hundreds of colleges, universities, churches and other institutions. He also gave talks and mathematical demonstrations on television and gave some lectures in the UK on his way back to India, in May 1958.

Bharati Krsna wrote sixteen volumes on Vedic Mathematics, one on each Sutra, but the manuscripts were irretrievably lost. He said that he would rewrite them from memory but owing to ill-health and failing eyesight got no further than writing a book intended as an introduction to the sixteen volumes. That book "Vedic Mathematics" written with the aid of an amanuensis is currently available and is the only surviving work on mathematics by this most remarkable man.

The ancient system of Vedic Mathematics was rediscovered from the Vedas between 1911 and 1918 by Sri Bharati Krsna Tirthaji (1884-1960). The word 'Veda' means 'knowledge' and also refers to a set of ancient Indian texts written in Sanskrit. These texts were written down about 1700 or 1600 BC according to western scholars but the content of the Vedas was passed on by an oral tradition long before writing was invented. The Vedas are said to cover every aspect and area of knowledge: grammar, architecture, ethics, astronomy etc. are all covered. During the nineteenth century there was a great interest in the Sanskrit texts in Europe. Bharati Krsna tells us some scholars ridiculed certain texts which were headed 'Ganita Sutras'- which means mathematics. They could find no mathematics in the translation and dismissed the texts as rubbish. Bharati Krsna, who was himself a scholar of Sanskrit, Mathematics, History and Philosophy, studied these texts and after lengthy and careful investigation was able to reconstruct the mathematics of the Vedas. According to his research all of mathematics is based on sixteen Sutras, or word-formulae. For example, 'Vertically and Crosswise` is one of these Sutras.

Bharati Krsna wrote sixteen volumes expounding the Vedic system but these were unaccountably lost and when the loss was confirmed in his final years he wrote a single book: Vedic Mathematics, currently available. It was published in 1965, five years after his death.

Today Vedic Mathematics is taught widely in schools in India and a great deal of research is being done. The first Vedic Mathematics Conference was in India in 1985.

Maharishi Mahesh Yogi began to explain the significance and marvellous qualities of Vedic Mathematics in 1988. He sent Dr Narinda Puri of Roorkee University on a world tour to promote Vedic Mathematics (mathematics with smiles). Maharishi schools around the world began to teach it. At the school in Skelmersdale, Lancashire, England a full course was written and trialled for 11 to 14 year old pupils, called The Cosmic Computer. This has proved very successful and is also to be published in the USA.

Interest in the Vedic system is growing in education where mathematics teachers are looking for something better and finding the Vedic system is the answer. Research is being carried out in many areas including the effects of learning Vedic Maths on children; developing new, powerful applications of the Vedic Sutras in geometry, calculus, computing etc.

VEDIC MATHEMATICS SUTRAS

1	एकाधिकेन पूर्वेन Ekādhikena Pūrvena	*By One More than the One Before*
2	निखिलं नवतश्चरमं दशतः Nikhilam Navataścaramam Daśatah	*All from 9 and the Last from 10*
3	ऊर्ध्वतिर्यग्भ्यामं Ūrdhva Tiryagbhyām	*Vertically and Cross-wise*
4	परावर्त्य योजयेत् Parāvartya Yojayet	*Transpose and Apply*
5	शून्यं साम्यसमुच्चये Sūnyam Sāmyasamuccaye	*If the Samuccaya is the Same it is Zero*
6	आनुरूप्ये शून्यं अन्यत् (Ānurūpye) Sūnyamanyat	*If One is in Ratio the Other is Zero*
7	संकलन व्यवकलनाभ्यां Saṅkalana Vyavakalanābhyām	*By Addition and by Subtraction*
8	पूरणापूरणाभ्यां Pūranāpūranābhyām	*By the Completion or Non-Completion*
9	चलनकलनाभ्याम् Calana Kalanābhyām	*Differential Calculus*
10	यावदूनं Yāvadūnam	*By the Deficiency*
11	व्यष्टिसमष्टिः Vyastisamastih	*Specific and General*
12	शेषाण्यङ्केन चरमेण Sesānyaṅkena Caramena	*The Remainders by the Last Digit*
13	सोपान्त्यद्वयमन्त्यं Sopāntyadvayamantyam	*The Ultimate and Twice the Penultimate*
14	एकन्यूनेन पूर्वन Ekanyunena Purvena	*By One Less than the One Before*
15	गुणितसमुच्चयः Gunitasamuccayah	*The Product of the Sum*
16	गुणकसमुच्चयः Gunakasamuccayah	*All the Multipliers*

SUB-SUTRAS

1	आनुरूप्येण Ānurūpyena	*Proportionately*
2	शिष्यते शेषसंज्ञः Śiṣyate Śeṣasamjñah	*The Remainder Remains Constant*
3	आधमाधेनान्त्यमन्त्येन Ādyamādyenāntyamantyena	*The First by the First and the Last by the Last*
4	केवलैः सप्तकं गुण्यात् Kevalaih Saptakam Gunyāt	*For 7 the Multiplicand is 143*
5	वेष्टनम् Vestanam	*By Osculation*
6	यावदूनं तावदुनं Yāvadūnam Tāvadūnam	*Lessen by the Deficiency*
7	यावदूनं तावदूनीकृत्य वर्गं च योजयेत् Yāvadūnam Tāvadūnīkrtya Vargañca Yojayet	*Whatever the Deficiency lessen by that amount and set up the Square of the Deficiency*
8	अन्त्ययोर्दशकेऽपि Antyayordaśake'pi	*Last Totalling 10*
9	अन्त्ययोरेव Antyayoreva	*Only the Last Terms*
10	समुच्चयगुणितः Samuccayagunitah	*The Sum of the Products*
11	लोपनस्थापनाभ्यां Lopanasthāpanābhyām	*By Alternate Elimination and Retention*
12	विलोकनं Vilokanam	*By Mere Observation*
13	गुणितसमच्चुयः समुच्चयगुणितः Gunitasamuccayah Samuccayagunitah	*The Product of the Sum is the Sum of the Products*
14	ध्वजाङ्क Dhvajānka	*On the Flag*

CHAPTER 1

- ODDS and EVENS
- COMPLEMENTS

HOW DO I START TEACHING MY CHILD THE FIRST PRINCIPLES OF VEDIC MATHEMATICS

The saying goes : "Any good children's book is always a good adult's book". To get started, for age 4 or 5 or 6, acquire a pouch of 25 pebbles (as 1+3+5+7+9 = 25 = 5 squared).

The ancients taught Mathematics through song, reciting the long unending decimals of pi (3·14159265...) as a hymn to Krishna or Shankacharya. Similarly, you can sing "The Times Table" and incorporate dance steps into it. You will be required also to make up and sing some ditties like : "5 Kookaburras on the clothes-line, 2 flew away, how many remain ?" Use your pebbles. Let the Child work it out through Discovery, Joy and Play.

Thus before we can do any calculation, the Child needs to understand the following 3 Principles :

1) ODDS and EVENS
2) COMPLEMENTS
3) DIGITAL SUMS

"Vedic Mathematics is the science and technology of precision and order, which maintains the integrity of Unity and at the same time spreads diversity within it".

by Maharishi Mahesh Yogi (Founder of Transcendental Meditation).

Maharishi's intention is to awaken global consciousness via the education of the 16 Sutras.

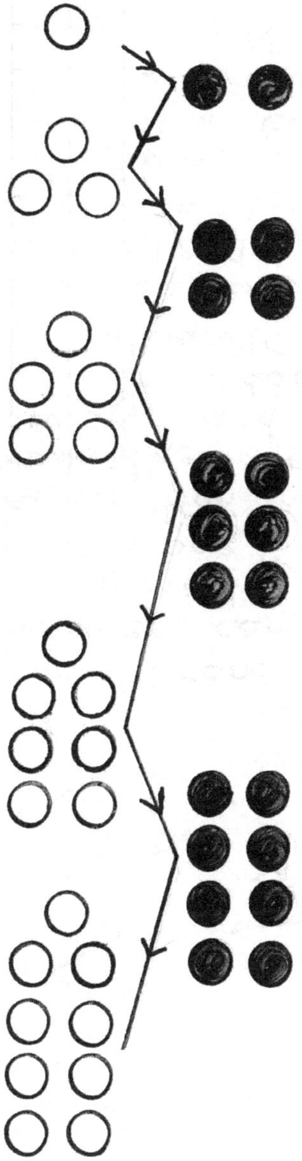

Use your pebbles or marbles to introduce the concept of Odd and Even numbers.

Examine the shapes thus formed.

Ask questions like : Q): "what happens if I add an odd no. to an even no.? eg: if I add 3 + 4 ?
Q): what happens if I add an odd no. + an odd no. ? eg 1+3 or 3+5 ?
Q): what happens if I add an even no. and an even no. ?

(no. = Number).

ON THE NATURE OF THE ODD AND EVEN NUMBERS

Perhaps a better and simpler way to illustrate The Odds and The Evens is :

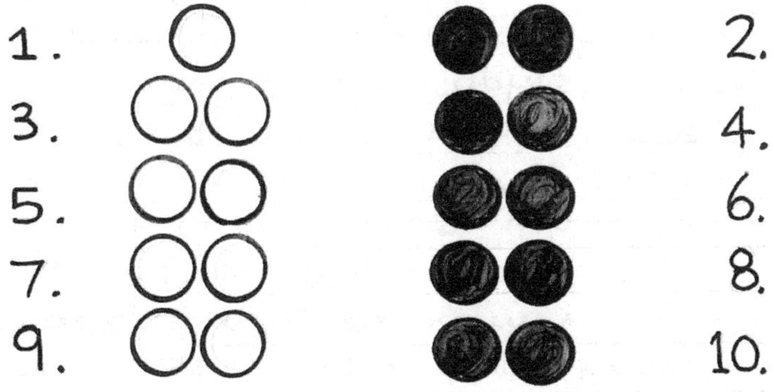

If using disks, you may choose to colour The Odds yellow, and The Evens purple, (they are complementary or opposite colours on the Colour Wheel).

A Child will first recognise THE SHAPE of 5 as ⁂ which is more meaningful than "5". At this stage, writing the number "5" is meaningless. It is your INTENTION to teach The Child PATTERN RECOGNITION.

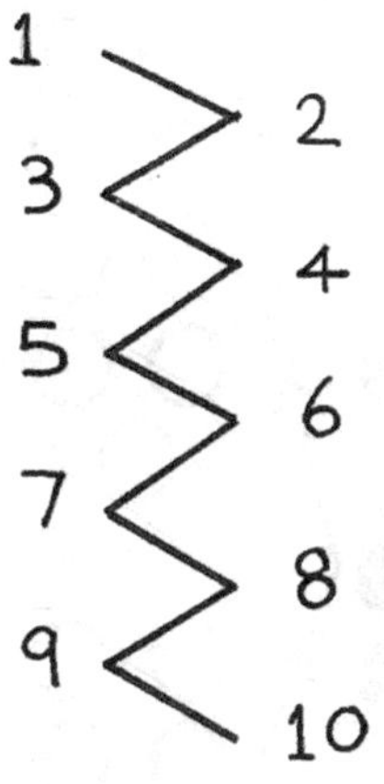

Explore counting numbers with this LADDER – LIGHTNING-FLASH diagram.

eg: to add 4 to 3 follow the arrows.

The symbol of the Lightning Bolt is akin to how the neurons in the brain fire. Answers can be arrived at "in a flash" in your head, or "in one line" on paper, rather than the current awkward down-the-page method of long division and multiplication.

Try adding 8 + 4.

nb: after 10 you go back up the Ladder to 1 (which is really 10) and continue: 8 + 4 = 12.

From this, we learn that adding an Even No. to an Even No. gives us another Even No. No real change.

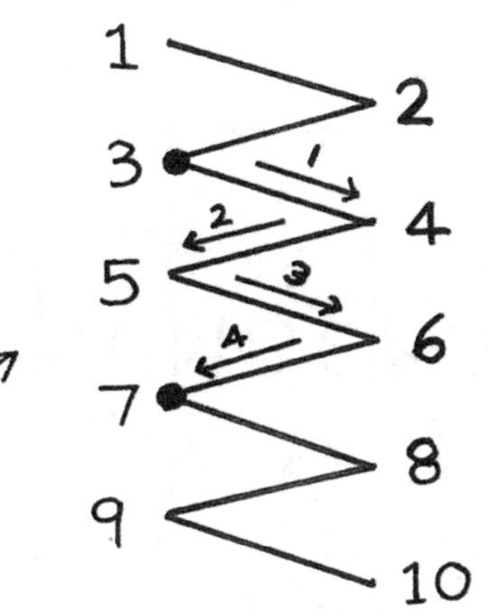

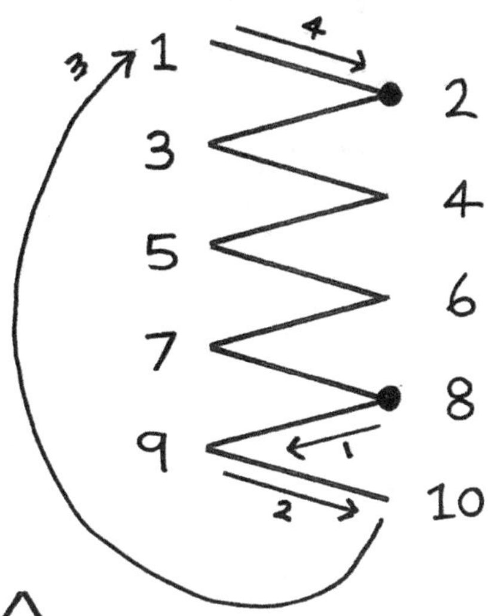

But observe what happens when we just add 1 to an Odd or an Even No. The Child will assess that the Energy of adding The Odd No. 1 to another number changes its Identity:

eg: an Odd No. + 1 = Even
or an Even No. + 1 = Odd.

This gives "1" a certain personality as the King of the Odds. It is for this reason that Pythagorean tradition has assigned Odds as male Nos. and Evens as Female.

(Later in life, a Child will learn the hidden knowledge contained within the nature of Magic Squares, our future codes, whether they be Odd or Even eg: 3×3 or 4×4 is critical in creating intelligent, sophisticated pollution-free motors to run Free-Energy Crafts as designed by Prof. J. R. Searl, the inventor of ...

pto

Earth-based UFO flying machines. (Refer to "The Book of Magic Squares" Vol.2. Jain).

Regarding the 'personality of numbers' Pythagoras (500 BC) and the Tibetan priests believe that numbers are sacred and have a Consciousness and therefore a Soul. Thus when they create Magic Squares, to heal the sick, by actually eating the rice wafers that it is written upon, they have in effect aligned a spiritual community of Beings. A Magic Square of 3×3 traditionally has

- all its columns = to 15
- all its rows = to 15
- its 2 diagonals = to 15
- 4 outer pairs of nos. equal double the central cell = 10
- The 4 outer corner nos. which are Even = 20
- The 4 odd nos. in a cross-shape also = 20.

6	1	8
7	5	3
2	9	4

CHAPTER 1

ODDS and EVENS

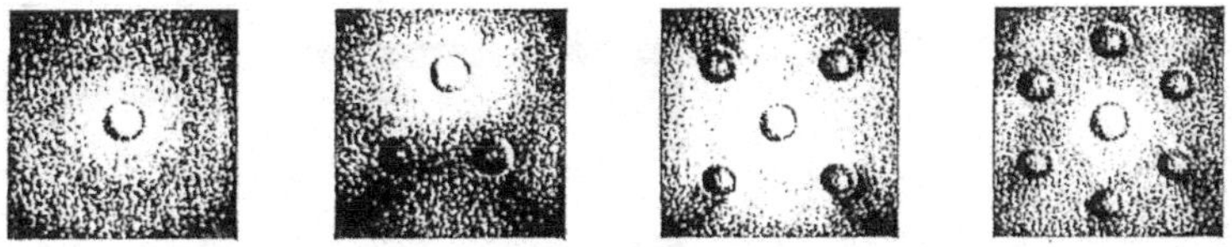

THE ODD (MASCULINE) NUMBERS

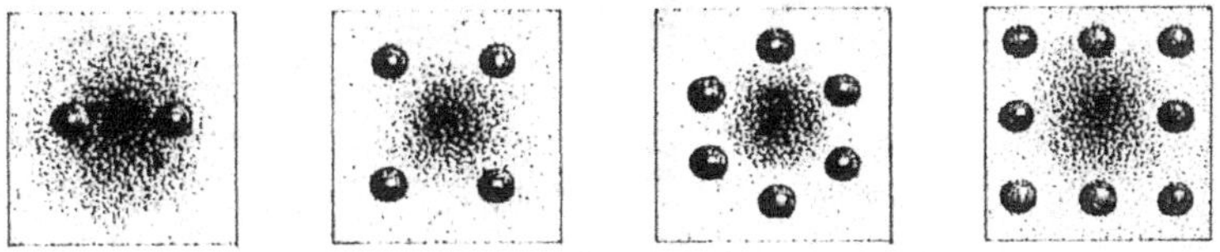

THE EVEN (FEMININE) NUMBERS

This unique nature of the Number 1, the ability to change an Odd to an Even no. or an Even to an Odd no. expresses its full power or prowess in its application of the first Sutra:
"BY ONE MORE THAN THE PREVIOUS DIGIT".

The 16 Sutras are really functions of The Mind eg: some can be used from Left to Right or Right to Left

o⟶ or ⟵o

ie. ambidextrously or VERTICALLY and CROSS-WISE

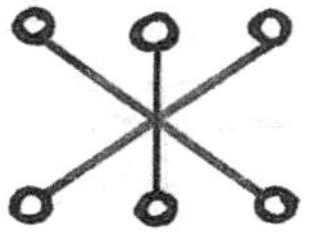

EXERCISE 1.1

ODD NUMBERS PATTERN RECOGNITION

The Sutra that is at work here is the first One: known as "BY ONE MORE THAN THE PREVIOUS DIGIT"

eg:	1	=	1 × 1
	1+2+1	=	2 × 2
	1+2+3+2+1	=	3 × 3
	1+2+3+4+3+2+1	=	4 × 4

Using your Pouch of 25 Pebbles, what is the common SHAPE of the Answers to the following Sums of the Odd Numbers a. → e.

Knowing this Shape, mentally predict the answers to the questions f, g, h, i, j, k and l.

What is your Conclusion?

a.	1	=	1	=	1 × 1	=	1^2
b.	1 + 3	=	4	=	2 × 2	=	2^2
c.	1 + 3 + 5	=	___	=	3 × ___	=	___
d.	1 + 3 + 5 + 7	=	___	=	___ × ___	=	___
e.	1 + 3 + 5 + 7 + 9	=	___	=	___ × ___	=	___
f.	1 + 3 + 5 + 7 + 9 + 11	=	___	=	___ × ___	=	___
g.	1 + 3 + 5 + 7 + 9 + 11 + 13	=	___	=	___ × ___	=	___
h.	1 + 3 + 5 + 7 + 9 + 11 + 13 + 15			=	___ × ___	=	___
i.	1 + ⟶ + 17			=	___ × ___	=	___
j.	1 + ⟶ + 19			=	___ × ___	=	___
k.	1 + ⟶ + 21			=	___ × ___	=	___
l.	1 + ⟶ + 23			=	___ × ___	=	___

EXERCISE 1.2

In Exercise 1.1, we saw how the Odd Nos. generated the Squared Nos.
To move into the next Dimension, with this knowledge, is to know how to go from The Family of Squares into The Family of Cubes, and using only ODD NUMBERS !
Add these sums :

a. 1 .... = $\underline{1 \times 1 \times 1}$ = $\underline{1^3}$

b. 3 + 5 .... = $\underline{2 \times 2 \times 2}$ = $\underline{2^3}$

c. 7 + 9 + 11 .... = $\underline{3 \times \quad \times \quad}$ = ___

d. 13 + 15 + 17 + 19 .... = ________ = ___

e. 21 + 23 + 25 + 27 + 29 ... = ________ = ___

f. 31 + 33 + 35 + 37 + 39 + 41 = ________ = ___

g. 43 + ⟶ + 55 = ________ = ___

ODD NUMBERS GENERATING CUBED NUMBERS

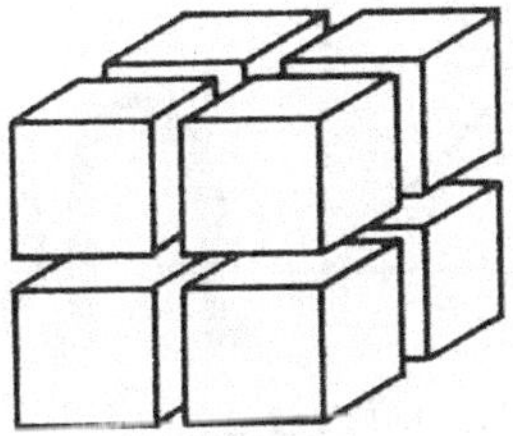

2 CUBED

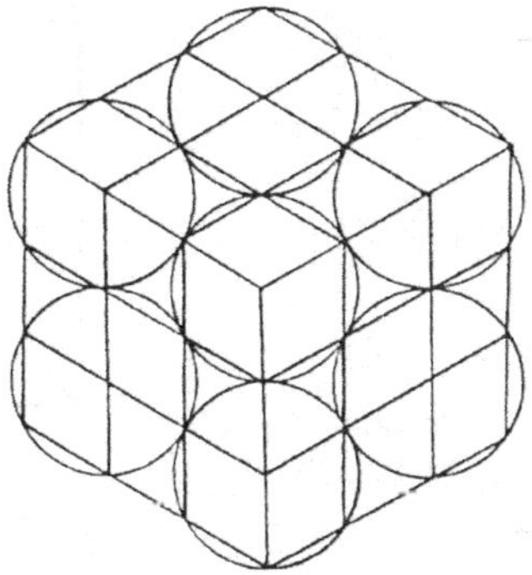

2 CUBED

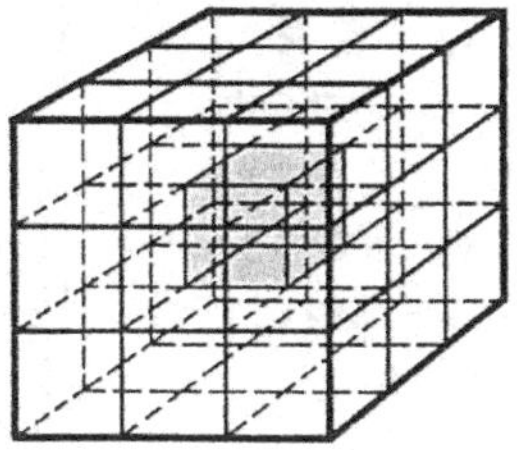

3 CUBED

4 SQUARE

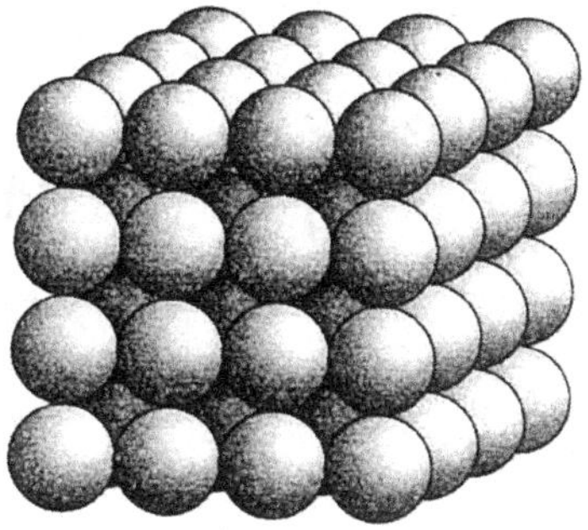

4 CUBED

Have you ever wondered about the origin of the numbers we use today ? where the actual hand-written script came from ?

A lot of people call our Western use of numbers as Arabic Numerals, but their origin goes further back to ancient Bharat or India !

Notice on your left that the Number 1 has 1 angle ! and the number 2 has 2 angles. Now you know why the 7 has a bar on its upright There are 7 angles.

Below is perhaps the geometric origin of all numerals : the "URDHVA TIRYAK" Sutra translated as : "Vertically and Crosswise".

AL-ARQUAN-AL-HINDU

"The HINDU FIGURES"

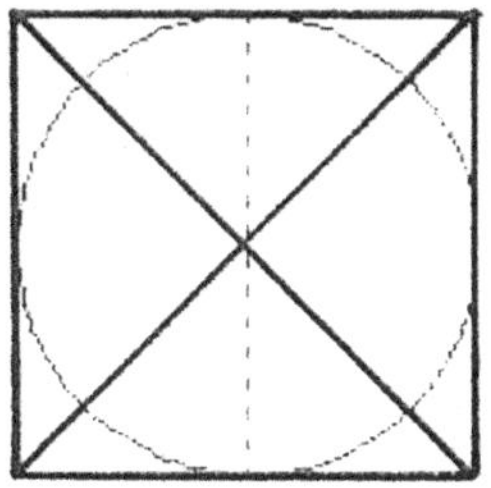

The Child is encouraged to understand that a number like 3 is not just 3 ! The intrinsic nature of 3ness actually consists of 4 Number Pairs or Number Partners. 3 seeds 4. Its components are not 3 but "One more than 3" ie 4. Thus again, the Sutra "BY ONE MORE" appears and underlies many mathematical principles).

0	and	3
1	and	2
2	and	1
3	and	0

By definition, a COMPLEMENT is : "that which completes or make perfect. Either of two parts or things needed to complete the whole".

Test your Child on all the Number Complements shown in this Chart. Teach them through pebble games or songs to memorise all the Number Pairs. eg :

COMPLEMENTS

1		2		3		4		5		6		7		8		9		10	
0	1	0	2	0	3	0	4	0	5	0	6	0	7	0	8	0	9	0	10
1	0	1	1	1	2	1	3	1	4	1	5	1	6	1	7	1	8	1	9
		2	0	2	1	2	2	2	3	2	4	2	5	2	6	2	7	2	8
				3	0	3	1	3	2	3	3	3	4	3	5	3	6	3	7
						4	0	4	1	4	2	4	3	4	4	4	5	4	6
								5	0	5	1	5	2	5	3	5	4	5	5
										6	0	6	1	6	2	6	3	6	4
												7	0	7	1	7	2	7	3
														8	0	8	1	8	2
																9	0	9	1
																		10	0

- "8 Horses in the paddock. 5 galloped away. How many left ?"
- "9 Wombats on the hill. 3 crawled into their burrow. How many left ?"
- "7 Kookaburras on the clothesline. 4 flew away. How many left ?"

The first SECRET to mastering Vedic Mathematics is knowing THE COMPLEMENTS of 9 and 10

In fact, the most important 2 numbers to master are all the Number Pairs for 9 and 10.

Memorising these complements will increase mental agility especially when solving problems like : 8976 x 9998.

This can be solved mentally by knowing the Nikhilam Sutra: "ALL FROM 9 AND THE LAST FROM 10". By knowing the complements of 9 and 10 beforehand, we mentally subtract all the digits from 9 and the last from 10 :

= 9 − 8 / 9 − 9 / 9 − 7 / 10 − 6

= 1 0 2 4 as the deficiency from 10,000. The whole setting out appears like this:

8 9 7 6 − 1 0 2 4

9 9 9 8 − 0 0 0 2

---

8 9 7 4 / 2 0 4 8

(Subtract 0002 from 8976 and multiply the deficiencies).

= 89,742,048 . Easy .

(This will be dealt with in-full, in Book 2).

The secret of Vedic Maths is understanding its system of unification and wholeness structured upon the Bases of 10, 100, 1,000, 10,000 etc.

Base 10 is our current Earth-based System, whereas Base 12 holds Galactic Memory. This will be revealed to you in the next chapter **3.7** called Digital Sums.

The original Base 10 Vedic System of mental arithmetic came under the body of knowledge called Engineering and Temple Building. It was called Sthāpathyaveda, also embracing architecture, structural human endeavour and all visual arts. Thus if a builder needed to subtract 784 from 1,000, he would apply the Sutra: "All From 9 and The Last From 10", and say :

9 − 7 / 9 − 8 / 10 − 4 and instantly know it was 216 !

EXERCISE 1.3

eg: 67 from 100 = 9−6 / 10−7
= 3 / 3

- Try these yourself, mentally using "All From 9 and the Last From 10".

a. 76 from 100
b. 888 from 1,000
c. 7,894 from 10,000
d. 67,896 from 100,000
e. 798,677 from 1,000,000

We use Complements all the time without even knowing we are using them! When you tell the time and say: its 6·45 you are referring to the ancient Babylonian system that was based on 60.
eg 15 and 45 = 60
30 and 30 = 60
and there are 60 minutes to each hour and 60 seconds to each minute.

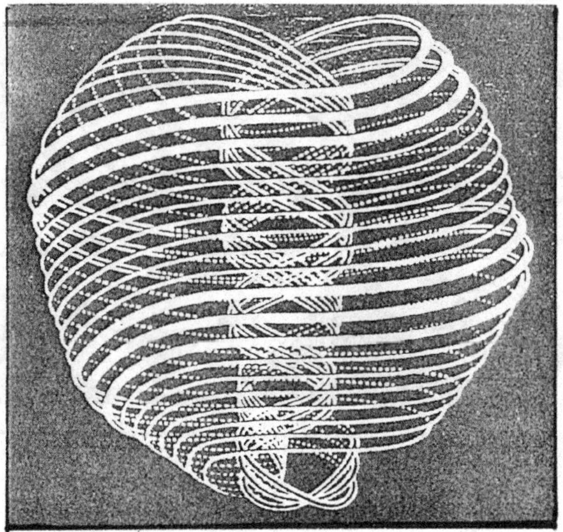
POSITIVE

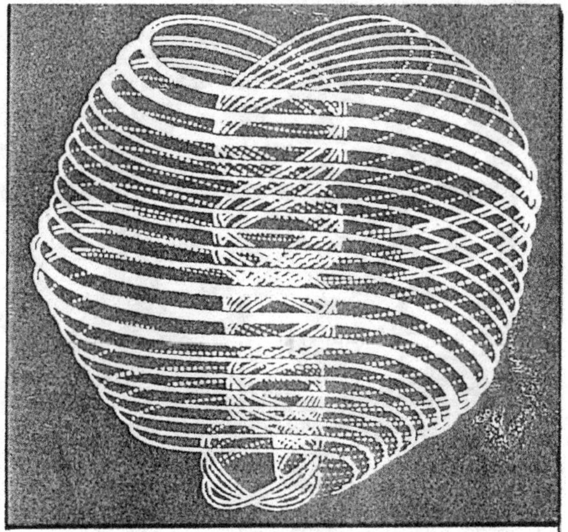
NEGATIVE

Depicted above is The ANU, the ultimate physical atom on the atomic sub-plane. The direction of their whorls are Complementary forces which pours through them. The 'positive' is the male, like a spring from which water bubbles out, and the 'negative' is the female, like a hole into which water disappears. (Reprinted from "Occult Chemistry" by Charles Leadbeater, a clairvoyant scientist, about 100 years ago.

For more info on Complements, especially the Chart for 11−20 Complements, refer to my book: "THE MAGIC OF NINE IN VEDIC MATHEMATICS" JAIN, 2001, PAGES 75 → 82.

Digit Sums

Chapter 2

DIGITAL SUMS

CHAPTER 2

INVESTIGATING

NUMBERS

The Hindus believe there are only 9 whole and primal numbers ie from 1 → 9.

When the next number 10 appears, they see it as 'the same and a dimension higher' than the original '1'. This Ladder of 1 → 9 is the Divine Template for all other numbers, for as you will see, all numbers can be 'reduced' back to 1 → 9.

Let us look at the next hyper-set of 1 → 9 known as the teen numbers, which when X-rayed, have a similar vibration:

1.	10	=	1 + 0	=	1
2.	11	=	1 + 1	=	2
3.	12	=	1 + 2	=	3
4.	13	=	1 + 3	=	4
5.	14	=	1 + 4	=	5
6.	15	=	1 + 5	=	6
7.	16	=	1 + 6	=	7
8.	17	=	1 + 7	=	8
9.	18	=	1 + 8	=	9

This act of reducing larger nos. by adding the sum of the digits (from the Latin word "digitus" means a finger or a toe), is called 'REDUCTION' or is better known as: DIGITAL SUMS.

Look at what happens when you keep adding the sum of all the 2-digit nos.:

10	→	18
19	→	27
28	→	36
37	→	45
46	→	54
55	→	63
64	→	72
73	→	81

It will be observed that the Series or Sequence of 9 counting nos. keeps repeating in this order of 1 → 9, for the higher nos., leading to the creation of The Multiplication Table (9 × 9). Notice that the numbers on the right is the 9 Times Table and is the key to lightning quick, mental calculations.

Any multi-digit no. can be reduced to a single digit merely by adding the sum of its digits.

eg 35 = 3 + 5 = 8

Sometimes it is necessary to add the digits again

eg 28 = 2 + 8 = 10 and

10 = 1 + 0 = 1

EXERCISE 2.1

Reduce the following numbers to a single digit:

a. 34
b. 67
c. 129
d. 1,247
e. 68,432
f. 100,003
g. 8,999,999

Digital Sums help us in

a) looking for patterns in nos.
b) in checking with large calculations to determine quickly that we have not erred. Rather than doing

"SAMUCCAYA" = DIGIT SUM

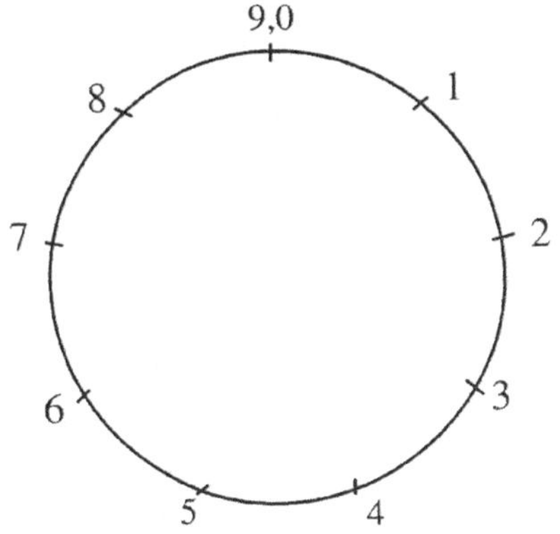

the 9-Point Circle is called the

ENNEAGRAM

the whole calculation of X, ÷, − or + again, we merely examine their Digital Sums. Such studies have evolved into certain Sutras like:

1) "The Sum of the Product is The Product of The Sum".
2) "By Completion or Non-Completion".

The very word "VEDA" means "The fountainhead and illimitable storehouse of all knowledge". The Vedas first introduced the Decimal System and the concept of Zero or SUNYA or BINDU, the origin of all creation; (from the latin word "decem" = Ten or "Decimus" = tenth).

Zero has 2 gifts:

1) it gives place value eg: 104 is differentiated from 1004. (The Babylonians and Roman Numerals lacked Zero).
2) it gives great power eg: 14 empowered by 100 gives 1400.

cont...

DIGITAL SUMS

ORIGIN of the WORD "CALCULATE" and ABACUS

(plural = Abaci)

Greek: "Abax" = tablet/board

Over the millenia, different cultures learnt to count, then add, subtract, multiply and divide. Some countries like Japan, China, and ancient Rome, made devices to deal easily with large computations. A special counting frame, shown below, which is still used today, was called an ABACUS, in which units, fives, tens, hundreds and thousands were represented by bead-heads that slid in grooves.

The name for these heads was CALCULI which is plural for CALCULUS or pebble, in the Latin language. An old name for lime is CALC, as is marble, another form of limestone, of which the Abacus heads were made of.

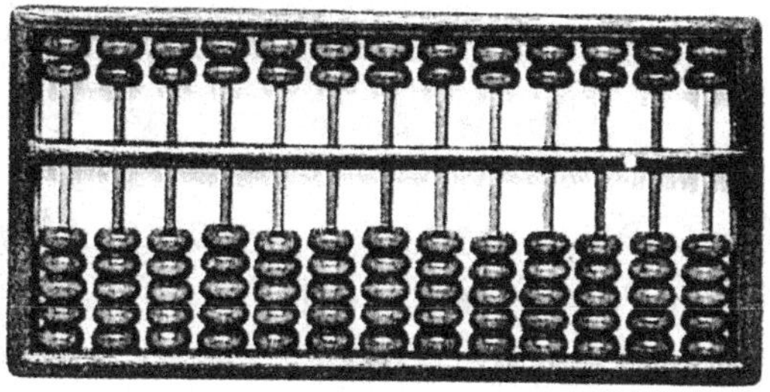

Some scholars say that Abacus is derived from a Semitic word "IBQ" meaning "to wipe the dust". The sand surface that was once used for writing developed into a board with lines, on which counters were used to depict numbers.

Some Chinese users of the Abacus associate it to the methodology of Vedic Mathematics and can use it faster than someone working with a calculator.

EXERCISE 2.2

The Brain loves and recognises Geometrical Patterns. It is for this reason that the 9 digits : 1, 2, 3, 4, 5, 6, 7, 8, 9 of our number system are appropriately displayed around a circle that has been divided into 9 lots of 40°.

This 9- Point Circle will be of great use when we examine the innate geometric harmony within the decimal nature of fractions like : $1 \div 7 = \cdot\dot{1}4285\dot{7}$

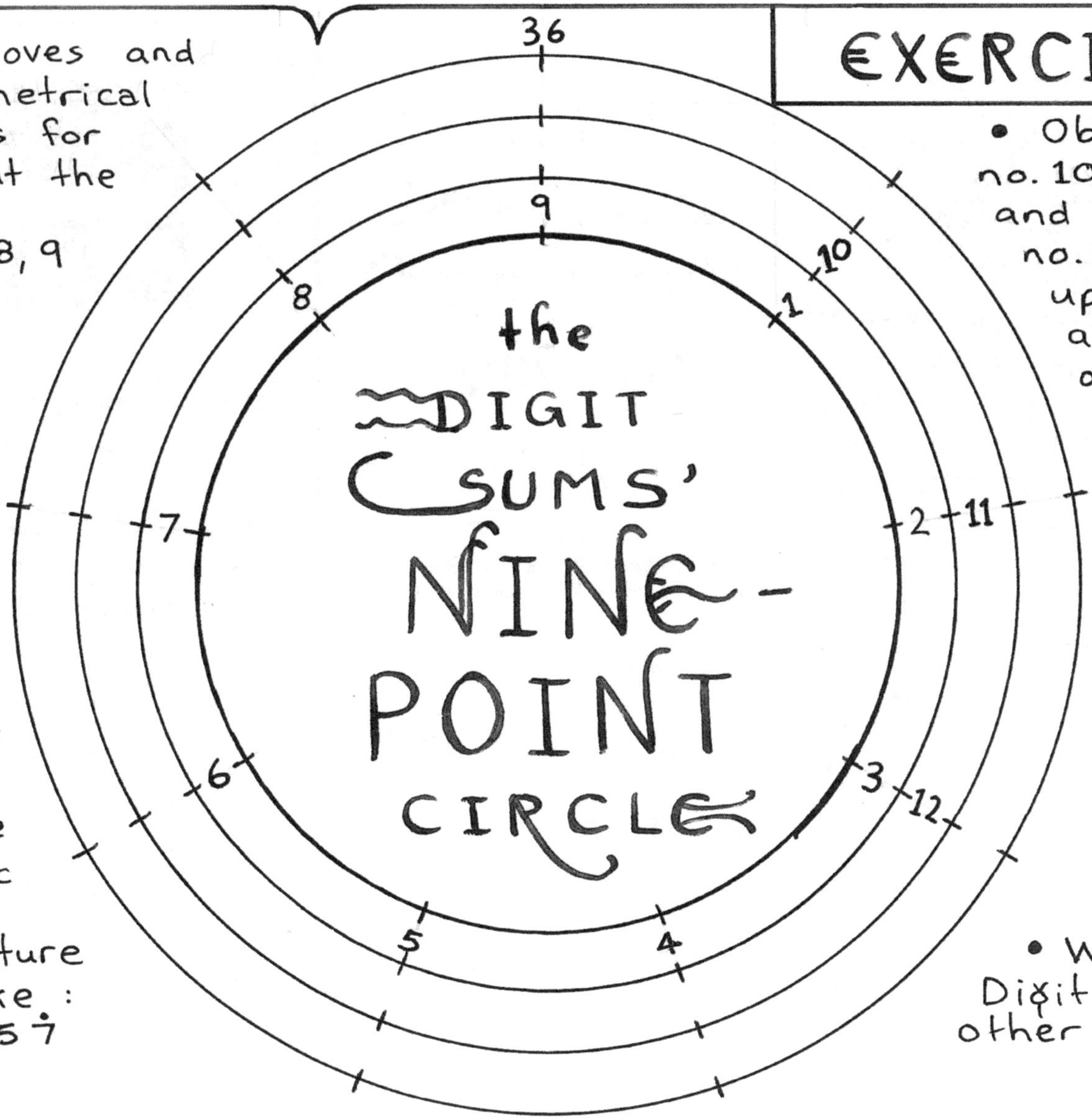

- Observe how the no. 10 jumped a ring and carries on till no. 12. Write down upon this diagram all the consecutive or natural counting numbers up to no. 36.

- Observe the Ray of numbers "1, 10, 19, 28" and add up their DIGIT SUMS ?
- Predict the next 5 nos. on this Ray.

- What are the Digit Sums for all the other Rays ?

This is achieved by repeated subtractions of 9 which leaves a Remainder. The Remainder is the Digit Sum.

eg: Cast Out The Nines from 33 ?

```
 33 -
  9
 ---
 24 -
  9
 ---
 15 -
  9
 ---
  6
```

As shown The Remainder is 6 which is the same as adding 3+3 (of 33).

A quick way, therefore, to find the Digit Sum of larger numbers is to "CAST OUT THE NINES"

eg: What is the Digit Sum of 912,399

Answer = 1 + 2 + 3 (by ignoring all the Nines) = 6

DIGITAL SUMS

"CASTING OUT THE NINES"

EXERCISE 2.3

Find the Digit Sums, by Casting Out The Nines, of the following :

a. 49	e. 93,199
b. 129	f. 192,009
c. 399	g. 329,911
d. 9,294	h. 15,229,990

nb: In the above example: f. 192,009, the Nines and the Zeroes are equivalent in their Digit Sums, so we only add 1 + 2 = **3**

nb: Another Digit-Sum Short-Cut, in the above example no. h. 15,229,990 is to cast out any group of numbers that also add up to 9. So we Cast out the Nines and also 5 + 2 + 2 = 9. This too is discarded and 1 is left, as the answer.

Chapter 2: THE DIGIT SUM CHECK:

"THE PRODUCT OF THE SUM IS THE SUM OF THE PRODUCTS"

EXERCISE 2.4

Since any group of numbers that add up to 9 can also be cast out, use this short-cut to calculate the following Digit Sums :

a. 725 d. 32,465
b. 524 e. 475,216
c. 9,336 f. 7,235,904

The PRODUCT of the SUM IS
The SUM of the PRODUCTS

To verify a complex multiplication you can use "Casting Out The Nines" :
eg Verify that
$2345 \times 6789 = 15,920,205$

- 2345 has a Digit Sum of 5 when 'Casting Out The 9s'
- 6789 . Digit Sum = 3
- 15,920,205 . Digit Sum = 6
- Multiply the first two Digit Sums ie $5 \times 3 = 15 = 1+5 = 6$.

Since this '6' is identical to the Digit Sum of the original product, then the multiplication is correct.

This Digit Sum check is an important Vedic Sutra known in Sanskrit as :
GUNITASAMUCCAYAH
SAMUCCAYAGUNITAH
which means :
The PRODUCT of the SUM IS
The SUM of the PRODUCTS.

EXERCISE 2.5

Verify that :

a) $345 \times 678 = 233,910$.
b) Does $6751 \times 3694 = 24,938,194$?
c) Does $452 \times 63,274 = 28,599,838$?

DIGIT SUMS

CHAPTER 2

SPLITTING THE DIGITS AND ADDING THEIR SUM

Here is a clever short-cut that involves Digit Sums and develops number skills.

To multiply any 2-digit no. by 11, just split the digits apart and insert between them the sum of those 2 digits :

eg Multiply 33×11

$= 3 \quad (3+3) \quad 3$

$= 3 \quad 6 \quad 3$

eg Multiply 71×11

$= 7 \quad (7+1) \quad 1$

$= 781$

EXERCISE 2.6

Mentally Multiply by 11 the following numbers :

a.	21	e.	63
b.	35	f.	71
c.	42	g.	81
d.	52	h.	90

In the previous examples, the Digit Sums were all < 9 (Less Than Nine). When the Digit Sum is > 9 (Greater Than Nine) you will have a 2-Digit teen number. Leave the unit's digit between the 2 split digits and carry the 1 to the first digit of the answer.

eg Multiply 48 by 11

$= 4 \quad (4+8) \quad 8$

$= 4 \quad 12 \quad 8$

$= 4 \quad {}_{1}2 \quad 8$

$= 5 \quad 2 \quad 8$

EXERCISE 2.7

Mentally Multiply by 11 the following numbers :

a.	38	e.	66
b.	28	f.	79
c.	49	g.	87
d.	56	h.	99

cont ...

Multiplication by ELEVEN

ADD THE NEIGHBOUR

This is the secret to multiplying larger nos. by 11.

eg Multiply 1,234 by 11

1) Add a Zero in front and at the end of the number :
   = 0 1 2 3 4 0
2) Drop the last Digit or add the neighbour to the right of the last digit
   = 4 + 0 = 4
   written as 012340
   4
3) Now go to 3. Add the Neighbour on the right
   = 3 + 4 = 7 012340
   74
4) Go to 2. Add the Neighbour = 2+3 = 5 012340
   574
5) Go to 1.
   Add the Neighbour
   = 1 + 2 = 3 = 012340
   3574
6) Go to 0.
   Add the Neighbour
   = 0 + 1 = 1 = 012340
   13574

Thus 1234 × 11 = 13574

EXERCISE 2.8

Multiply by 11 :

a. 123	d. 53,024
b. 2,453	e. 24,513
c. 4,162	f. 112,358

Notice in the last example Ex 2.8f when 'Adding The Neighbour' gave a Digit Sum in the teens, which meant there was a carry-over to the next digit on the left:

112,358 × 11 = 0112358~~0~~

+1

1235938

Let us check that

12345 × 11 = 0123450
135795

It is good practice to have all your calculations checked by Digit Sums.

Presented here is another oddity.

1) Place a dot under every 2nd no. beginning from the right 135795 (dots under 1, 5, 9)

2) Add the digits dotted
1 + 5 + 9 = 15

3) Add the digits not dotted
3 + 7 + 5 = 15

4) The answer is correct when both digit sums are the same (ie 15 = 15)! Sometimes they differ by a multiple of 11 (this means the differences between the 2 Digit Sums can be 11, 22, 33, 44, 55 etc.

DIGIT SUM CHECK FOR MULTIPLICATION BY ELEVEN

or

(DIVISIBILITY BY 11)

EXERCISE 2.9

Check that the Digit Sums, for the following nos., prove DIVISIBILITY BY 11, by answering Yes or No.

a. 356,576
b. 38,731,781
c. 252,646
d. 986,194

EXERCISE 2.10

Check for DIVISIBILITY BY 11 for the Answers in Exercise 2.8

eg Ex 2.8f

112358 × 11 = 01123580
1235938 (dots under 2, 5, 3)

Add dotted digits = 2 + 5 + 3 = 10
Add undotted = 1 + 3 + 9 + 8 = 21
The difference between the two is 21 − 10 = 11
∴ Yes.

ANOTHER PERSPECTIVE × 11

Here is simply another view to understand numerically what "ADDING THE NEIGHBOUR" really does :

1234 × 11 =

```
1 2 3 4 +            + 1 2 3 4
  1 2 3 4            1 2 3 4
--------->           <---------
1 3 5 7 4            1 3 5 7 4
```

Notice how by sliding the 2nd Rows, whether to the Left or to the Right, achieves the same result.

So what do you think the result will be if you subtracted the 2nd Row from the 1st row by sliding 1 column to the Right: ?

```
=   1 2 3 4  _
      1 2 3 4
    --------->
```

DIGIT SUMS CHAPTER 2

MULTI-PLICATION BY 11

"ANOTHER PERSPECTIVE" + "PALINDROMES"

The answer is 11106 which is 1234 × 9 !

So here is a curious relationship between × 11 and × 9 . It is a clue to understanding the merging of 2 Vedic Sutras, when thinking in Base 10, that × 11 = "By One More Than The One Before" (or Ekādhikena Pūrvena) and that × 9 = "By One Less Than The One Before" (or Ekanyunena Purvena) .

EXERCISE 2.11

Multiply by 11, in this fashion, the following nos:

a. 1111
b. 1221
c. 123321
d. 12344321

Notice that the Answers are "PALINDROMIC" (Gk: 'Palindromos' = running back") having reflection about the central number. Notice how the DIGIT SUMS DOUBLE eg 1221 → 13431.

Chapter 3 : the Digit Sums of

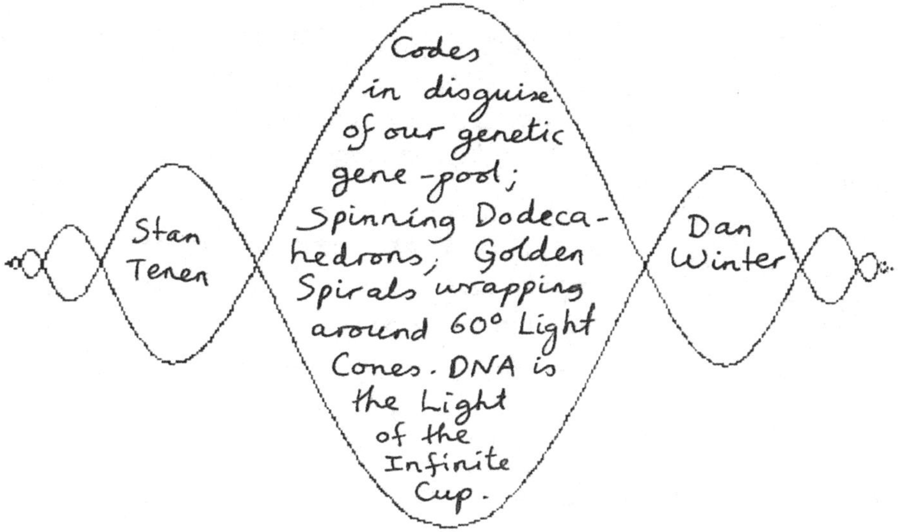

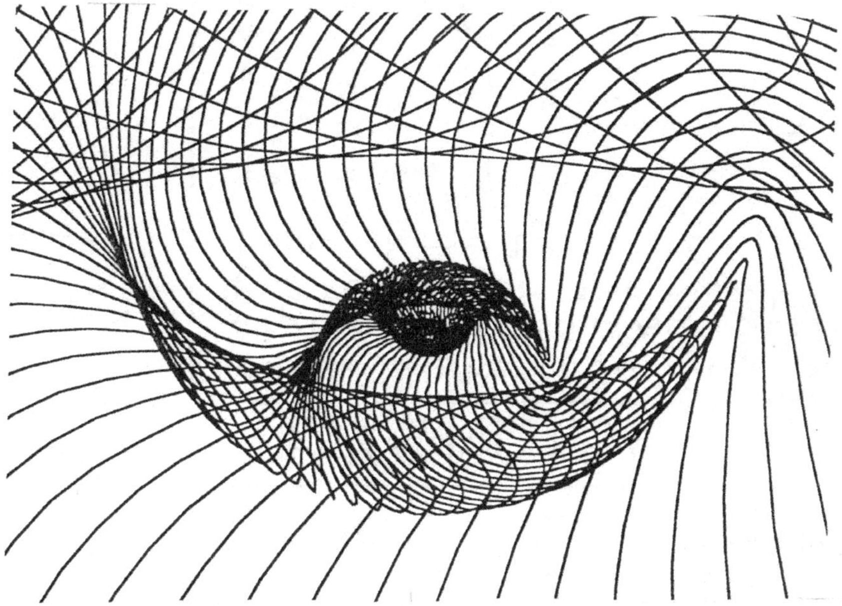

the Universal Number Series

DIGIT SUMS

CHAPTER 3

"UNIVERSAL NUMBER SERIES"

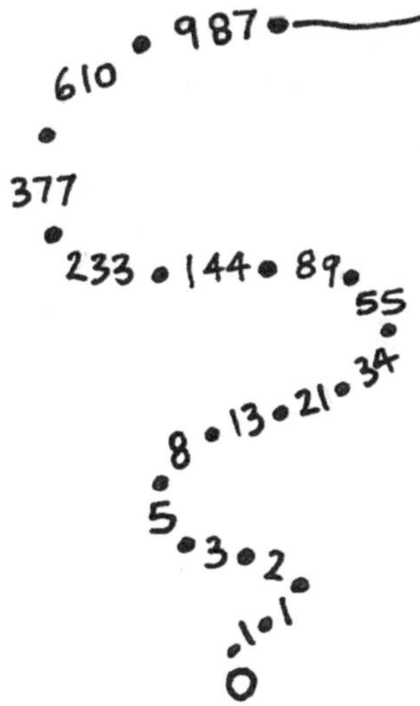

To grasp the fundamental truth of VM, it is important to be a PATTERN HUNTER.

This Chapter will explore Digit Sums of certain timeless and eternal Number Series, most of which you will be familiar with. Such terms are called CONSTANTS : ie they will never change, a forever existing galactic mathematics. They are the blueprints of your Genetic Code, past, now and yet to come.

To determine simple inherent patterns or encodings that lie hidden within the following Series of numbers, your only tool is to use the powerful X-ray scanning device called DIGITAL SUMS.

On another LEVƎ⅃, you actually know all this information. Allow yourself not to memorise, but to Re-Member all the following Series, all found in the Mathematics of Nature. They are part of your Ascension Process, if not for you, but for your Children.

Here is a simple example :

EXERCISE 3.1

What is the recurring or repeatable DIGITAL SUM PATTERN for the Numbers from 1 → 10 CUBED ?

1^3	=	1			=	1
2^3	=	8			=	8
3^3	=	27	=	2+7	=	9
4^3	=	64	=	6+4	=	1
5^3	=	125	=	1+2+5	=	8
6^3	=	216	=	2+1+6	=	9
7^3	=	343	=	3+4+3	=	1
8^3	=	512	=	5+1+2	=	8
9^3	=	729	=	7+2+9	=	9
10^3	=	1,000	=	1+0+0+0	=	1

The answer to Exercise 3.1 is shown as the far right-hand-side column of digits : 1, 8, 9 , 1, 8, 9 , 1, 8, 9 , 1

You can see at a glance that the recursion that is happening is based on the 3 Digits "1, 8, 9".
We therefore say that
a) The Digital Sum Series Pattern is 1, 8, 9
b) It has a Periodicity or Period of P = 3
It is derived from the Greek word "PERIODUS", meaning "a going around, cycle".

PERIODICITY of the TRIANGULAR NUMBER SERIES

EXERCISE 3.2

Here are the first 28 Δ TRIANGULAR NUMBERS.
Formula is : $[n \times (n+1)] \div 2$
eg $T5 = (5 \times 6) \div 2 = 15$
Find its Digital Sum Series Pattern and its Periodicity ?

T1 =	1	____	T19 =	190	____
T2 =	3	____	T20 =	210	____
T3 =	6	____	T21 =	231	____
T4 =	10	____	T22 =	253	____
T5 =	15	____	T23 =	276	____
T6 =	21	____	T24 =	300	____
T7 =	28	____	T25 =	325	____
T8 =	36	____	T26 =	351	____
T9 =	45	____	T27 =	378	____
T10 =	55	____	T28 =	406	____
T11 =	66	____			
T12 =	78	____			
T13 =	91	____			
T14 =	105	____			
T15 =	120	____			
T16 =	136	____			
T17 =	153	____			
T18 =	171	____			

Write the Digit Sum for each Δ no. on the corresponding line.

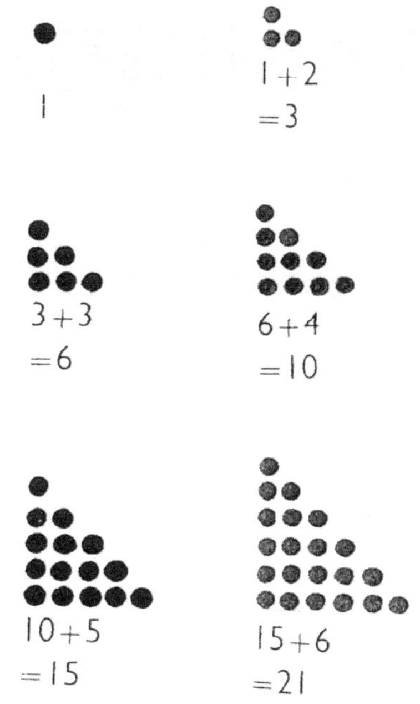

TRIANGULAR NUMBERS. Write in this box the Digital Sum Series Pattern and its Periodicity.

__ __ __ __ __ __ __ __ __

Periodicity (P) = ____

CHART OF THE SQUARES

NUMBER	SQUARED NUMBER	MAGIC SQ. CONSTANT	SUM OF THE NATURAL SQ.
1	1	1	1
2	4	5	10
3	9	15	45
4	16	34	136
5	25	65	325
6	36	111	666
7	49	175	1225
8	64	260	2080
9	81	369	3321
10	100	505	5050
11	121	671	7381
12	144	870	10440
13	169	1105	14365
14	196	1379	19306
15	225	1695	25425
16	256	2056	32896
17	289	2465	41905
18	324	2925	52650
19	361	3439	65341
20	400	4010	80200
21	441	4641	97461
22	484	5335	117370
23	529	6095	140185
24	576	6924	166176
25	625	7825	195625
26	676	8801	228826
27	729	9855	266085
28	784	10990	307720
29	841	12209	354061
30	900	13515	405450
31	961	14911	462241
32	1024	16400	524800
33	1089	17985	593505
34	1156	19669	668746
35	1225	21455	750925
36	1296	23346	840456
37	1369	25345	937765
38	1444	27455	1043290
39	1521	29679	1157481
40	1600	32020	1280800
41	1681	34481	1413721
42	1764	37065	1556730
43	1849	39775	1710325
44	1936	42614	1875016
45	2025	45585	2051325
46	2116	48691	2239786
47	2209	51935	2440945
48	2304	55320	2655360
49	2401	58849	2883601
50	2500	62525	3126250
51	2601	66351	3383901
52	2704	70330	3657160
53	2809	74465	3946645
54	2916	78759	4252986
55	3025	83215	4576825

NO 1

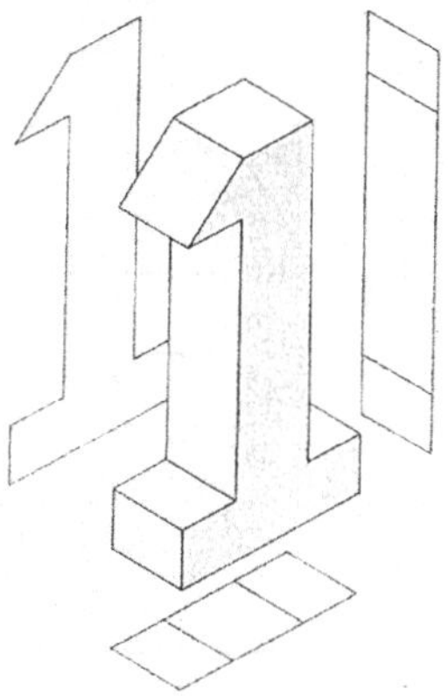

The CHART of The SQUARES will be the focus for the discovery of more DIGITAL SUM SERIES with unique PERIODS or PERIODICITIES: 1) Squared Numbers, which begins on this page, 2) The Last Digits of these Squared Numbers 3) The Magic Squares have CONSTANT or eternal harmonic Sums and 4) The Sum of the Natural Squares. Sometimes we need to explore a Series, beyond the first 10 Numbers as indicated below : pto

PYRAMID of ONES : DIGITAL SUMS

1 x 1 = 1	= 1	= 1	$= 1^2$
11 x 11 = 121	= 1+2+1	= 4	$= 2^2$
111 x 111 = 12321	= 1+2+3+2+1	= 9	$= 3^2$
1111 x 1111 = 1234321	= 1+2+3+4+3+2+1	= 16	$= 4^2$
11111 x 11111 = 123454321	=	= 25	$= 5^2$
111111 x 111111 = 12345654321	=	= 36	$= 6^2$
1111111 x 1111111 = 1234567654321	=	= 49	$= 7^2$
11111111 x 11111111 = 123456787654321 ...	=	= 64	$= 8^2$
111111111 x 111111111 = 12345678987654321	=	= 81	$= 9^2$
1111111111 x 1111111111 = 123456790087654321	=	= 100	$= 10^2$

DIGITAL SUM SERIES PATTERN for the a) SQUARED NUMBERS and b) LAST DIGITS of the SQUARED NUMBERS

Find the Digit Sum for the following 36 Squared Numbers in the box provided :

No. = Number
SQ. = Squared Number
D.S. = Digit Sum

No.	SQ.	D.S.	No.	SQ.	D.S.
1	1		19	361	
2	4		20	400	
3	9		21	441	
4	16		22	484	
5	25		23	529	
6	36		24	576	
7	49		25	625	
8	64		26	676	
9	81		27	729	
10	100		28	784	
11	121		29	841	
12	144		30	900	
13	169		31	961	
14	196		32	1024	
15	225		33	1089	
16	256		34	1156	
17	289		35	1225	
18	324		36	1296	

EXERCISE 3.3

After having written down the Digit Sum of the Squared Numbers determine the recurring Digital Sum Series Pattern and its Periodicity.

_ _ _ _ _ _ _ _ _ _ _

Periodicity (P) = ______

Write the 2 answers above.

EXERCISE 3.4

Observing the Chart of the first 36 Squared Numbers study only their Last Digits What is its Digital Sum Series Pattern and its Periodicity ?

_ _ _ _ _ _ _ _ _ _ _ _

Periodicity (P) = ______

A MAGIC SQUARE (M.SQ.) by definition, has the sum of its Columns, Rows and Diagonals all equal to the same sum, known as The MAGIC SQUARE CONSTANT.

No.	M.SQ.	D.S	No.	M.SQ.	D.S.
1	1		19	3439	
2	5		20	4010	
3	15		21	4641	
4	34		22	5335	
5	65		23	6095	
6	111		24	6924	
7	175		25	7825	
8	260		26	8801	
9	369		27	9855	
10	505		28	10990	
11	671		29	12209	
12	870		30	13515	
13	1105		31	14911	
14	1379		32	16400	
15	1695		33	17985	
16	2056		34	19669	
17	2465		35	21455	
18	2925		36	23346	

Digital Sums Series Patterns for the MAGIC SQUARE CONSTANTS

6	1	8
7	5	3
2	9	4

EXERCISE 3.5

The first 36 Magic Square Constants have been listed, in the table on your left, under "M.SQ.". In the columns marked "D.S" write in the Digit Sum of those 36 M.Sq. Constants and then determine the existence of a specific repetition. Once ascertained, write inside the box below, the Digital Sum Series Pattern, on the dashed lines, and also its Period.

_ _ _ _ _ _ _ _ _ _ _

Periodicity (P) = ____

This Digital Sum X-raying of data will prove useful in Vedic Mathematics later on, in your ability to understand Patterns.

A NATURAL SQUARE, by definition, is merely the natural counting numbers arranged into its square form. It is from this sequential or consecutive order that the numbers are intelligently juggled around to form a Magic Square.

No.	Σ NATURAL SQ.	D.S
1	1	
2	10	
3	45	
4	136	
5	325	
6	666	
7	1225	
8	2080	
9	3321	
10	5050	
11	7381	
12	10440	
13	14365	
14	19306	
15	25425	

DIGITAL SUMS SERIES PATTERN for the NATURAL SQUARES

1	2	3
4	5	6
7	8	9

EXERCISE 3.6

The Sum of all the digits shown in the Natural Square of 3×3 (bottom, centre of page) is :
$1+2+3+4+5+6+7+8+9 = 45.$

Another way of representing this is : $\sum_{1}^{9} = 45$

which reads as "SIGMA 1 to 9 = 45". SIGMA is the 18th letter of the Greek Alphabet and means "Sum Of".

On the Left Hand Side Table, in the Column marked "D.S" insert the Digit Sum of the 15 listed "Σ Natural Squares", and ascertain the D.S.S.P or Digital Sum Series Pattern + Periodicity, then write the answers below :

___ ___ ___

Periodicity (P) = ___

1·618033988 7

the FIBONACCI SEQUENCE

EXERCISE 3.7

Here are the first 24 numbers of the Fibonacci Series, the Living Maths of Nature. The Phi Ratio 1:1·618 is obtained by dividing terms like : 610 ÷ 377 .

1, 1, 2, 3, 5, 8, 13, 21, 34, 55, 89, 144, 233, 377, 610, 987, 1597, 2584, 4181, 6765, 10946, 17711, 28657 and 46368 . Find the D.S.S.P. or Digital Sum Series Pattern by writing the answer in the box below. The Periodicity = 24.

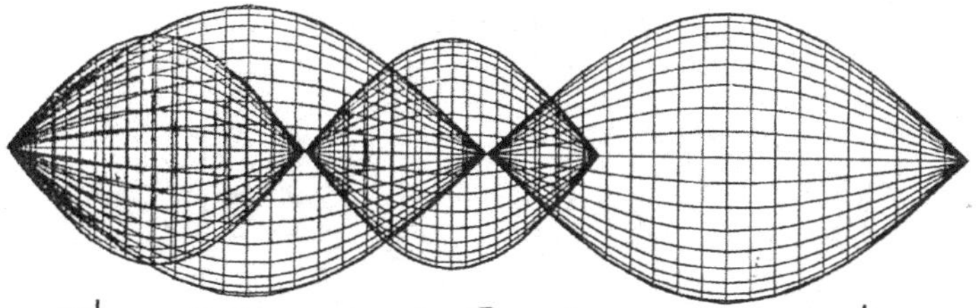

The Fibonacci Series is part of our DNA Genetic Encoding.

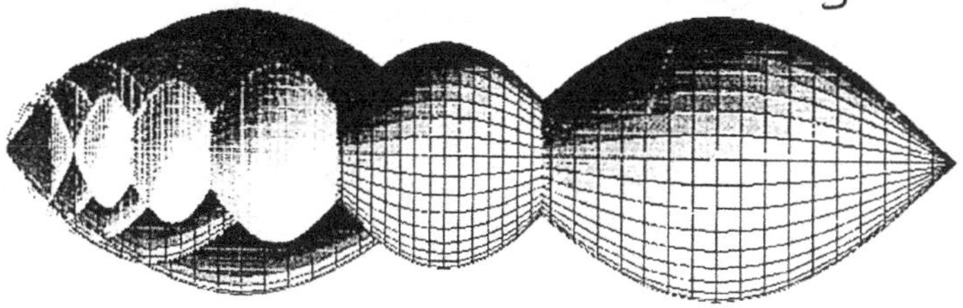

_ _ _ _ _ _ _ _ _ _ _ _

_ _ _ _ _ _ _ _ _ _ _ _

Look very closely at these 2 lines of Digit Sum numbers, and what do you notice ?

$33^2 = 1089$

$333^2 = 110889$

$3333^2 = 11108889$

$33333^2 = 1111088889$

$333333^2 = ?$

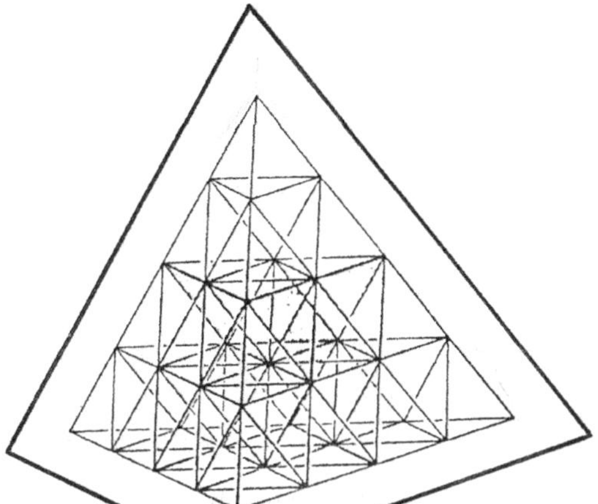

A PYRAMID of THREES

EXERCISE 3.8

Find the Digit Sums for this Pyramid of Threes. What is the Digital Sum Series Pattern and its Periodicity? Insert the answers inside the box below :

D.S.S.P = __________

Periodicity (P) = ____

As this is an easy task, you are challenged, without the use of a calculator, to determine what the answers are for the following 4 squared nos. by observing the obvious pattern :

$333,333^2$ = __________

$3,333,333^2$ = __________

$33,333,333^2$ = __________

$333,333,333^2$ = __________

EXERCISE 3.9

QUESTION 1 : What are the sum totals of each of the 8 Rows ? eg $\sum$ Row 4 = 1+3+3+1 = ?

Row 1 = ____ Row 5 = ____

Row 2 = ____ Row 6 = ____

Row 3 = ____ Row 7 = ____

Row 4 = ____ Row 8 = ____

QUESTION 2 : What do you notice about all the totals of the 8 rows ?

. . . .

. . . .

QUESTION 3 : By foreseeing or understanding the pattern, what are the sum totals (or "Σ" Sigma) for :

Row 9 = ____

Row 10 = ____

Row 11 = ____

Row 12 = ____

Row 13 = ____

PASCAL'S TRIANGLE or HALAYUDHA'S Δ

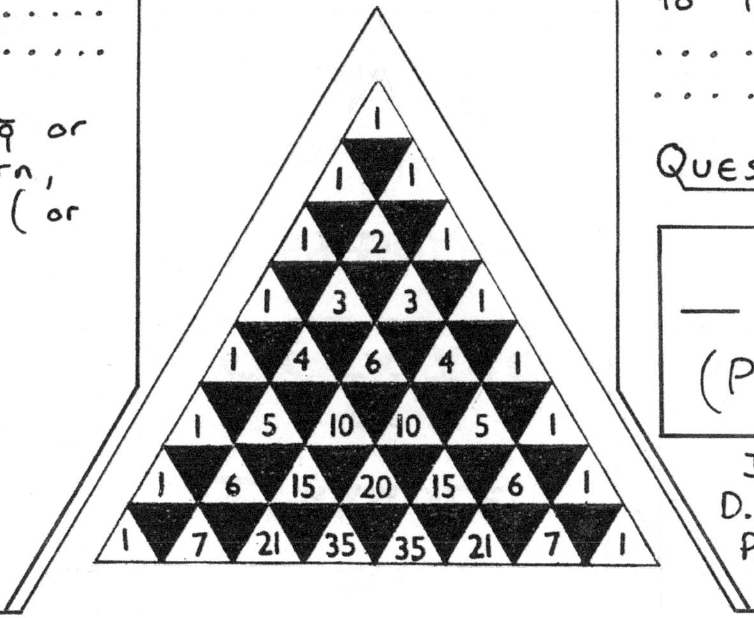

QUESTION 4 : What are the Digit Sums of the 13 established Rows ? Write the answers below :

__ __ __ __ __ __

__ __ __ __ __ __ __

QUESTION 5 : With the limited information of 8 Rows, why did we need to extend our data to 13 Rows ?

. . . .

. . . .

QUESTION 6 :

__ __ __ __ __ __
(P) = ____

Insert inside this box, the D.S.S.P (Digital Sum Series Pattern, and its Periodicity.

THE PARABLE of SESSA'S CHESS-BOARD (on the POWERS of 2)

Having discovered that the Rows of Pascal's / Halayudha's Triangle add up to the increasing Powers of 2, here is a beautiful and famous story about repeated doubling generating huge totals whose moral is : "It pays well to know your mathematics".

The inventor of The Game of Chess was Sessa, an Indian Mathematician considered to be a learned Pandit (or wise man). He taught this game to a King Purushottama of Persia (known today as Iran), who boasted, over the years, that he was an undefeated champion.

The story goes that Sessa succeeded in beating the proud King and offered Sessa whatever he wished for.

Sessa's apparently insignificant wish or reward was asking for 1 grain of rice (some stories say wheat) for the 1st square of the Chess-Board, 2 for the 2nd Square, 4 for the 3rd Square, 8 for the 4th square etc up to the 64th or final square.

The wish was granted and the King's wise men started doubling the numbers and arrived at a 20 digit answer :
18446744073709551615
grains of rice.

The King noticed that the sacks of rice were slow in coming and enquired from his trusted minister the reason why. They told him in one pint of rice there are 9,000 grains which approximate 32,000,000,000,000 bushels (or 32×10^{12}) which is many times the world's granaries annual rice production.

Thus Sessa became a very very rich man!

[RESEARCH : This 20 digit number $2^{64}-1$ appears in another famous legend. Find a good book and research "THE TOWER OF BRAHMA].

Tribute to Halayudha's "Powers of Eleven"

Blaise Pascal

"....it was the Jaina mathematician, Halayudha (3rd century BC) who first recognised the arithmetic triangle for which Pascal is given the credit. (It is better called the Meru-Prastera rule and was discovered about 13 centuries before Pascal.) It has to do with the coefficients of the successive terms in the expansion of the binomial expression $(a + b)^n$. It is also basic in the calculation of probabilities."
Number by John McLeish 1991 ISBN 0 7475 0921 2.

Pascals Triangle usually looks like this:-

1
1 1
1 2 1
1 3 3 1
1 4 6 4 1
1 5 10 10 5 1
1 6 15 20 15 6 1
1 7 21 35 35 21 7 1
1 8 28 56 70 56 28 8 1
1 9 36 84 126 126 84 36 9 1
1 10 45 120 210 252 210 120 45 10 1
1 11 55 165 330 462 462 330 165 55 11 1

And so on.

The triangle appears to be a presentation of numbers in the ratio 1:11 starting with 1, so that each row of numbers is a power of 11. Each new line is being calculated by multiplying the previous line by 11, ie. the first and last digits of the previous number are put down as they are, and the middle numbers are each pair of numbers added together. (We can multiply any number by 11 using this method).

CONTINUED OVERLEAF

PASCAL'S/ HALAYUDHA'S TRIANGLE as POWERS of ELEVEN

DID YOU KNOW?

that each row of Pascal's Triangle is really a Power of 11 (as well as a Power of 2 in its Digit Sums). Observe :

- Row 1 = 1 = 11^0
- Row 2 = 11 (or 11 × 1) = 11^1
- Row 3 = 121 (or 11 × 11) = 11^2
- Row 4 = 1331 (or 11 × 11 × 11) = 11^3
- Row 5 = 14641 = 11^4
- Row 6 = 1 5 10 10 5 1

Observe the Double Digits (10). To get the correct Power of Eleven, we need to process the "carry-overs"

ie = 1 5 ₁0 ₁0 5 1 (carry +)

= 1 6 1 0 5 1

= 161,051 = 11^5

EXERCISE 3.10

- What is Eleven to the 6th Power ie. 11^6 ?

Clue. Write down the 7th Row of Pascal's △ and process the Carry-Overs :

= 1 6 15 20 15 6 1 (carry +)

= _ _ _ _ _ _ _

- What is Eleven to the 7th Power ie 11^7 ?

= 1 7 21 35 35 21 7 1

= _ _ _ _ _ _ _ _

- What is Eleven to the 8th Power ie 11^8 ?

= 1 8 28 56 70 56 28 8 1

= _ _ _ _ _ _ _ _ _

Here is the traditional 9x9 Multiplication Table

By finding the Digit Sums of all the double-digited numbers (in Exercise 3.11) you will have generated another 9 rows of universal data that each constitute another Digital Sum Series Pattern explored over the next 13 pages :

X	1	2	3	4	5	6	7	8	9
1	1	2	3	4	5	6	7	8	9
2	2	4	6	8	10	12	14	16	18
3	3	6	9	12	15	18	21	24	27
4	4	8	12	16	20	24	28	32	36
5	5	10	15	20	25	30	35	40	45
6	6	12	18	24	30	36	42	48	54
7	7	14	21	28	35	42	49	56	63
8	8	16	24	32	40	48	56	64	72
9	9	18	27	36	45	54	63	72	81

HOW TO CREATE THE VEDIC SQUARE

EXERCISE 3.11

Convert all the double digit numbers to their respective Digit Sums, in the vacant box provided below:

eg 64 = 6+4 = 10 = 1+0 = 1

(Single digits of the 9x9 Multiplication Table remain the same).

EXERCISE 3.12

Blacken all the vacant cells below that correspond to all the Cells or Points or Digits of 1 (ness), in the Left box. Similarly, in the Right box, blacken all the vacant Cells of 8 (ness).

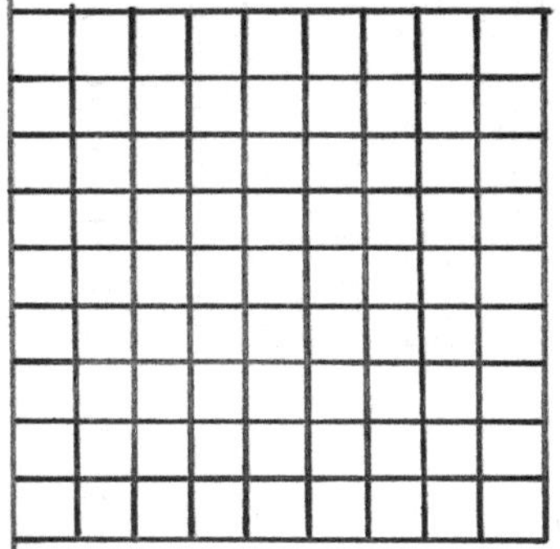

ALL THE ONES

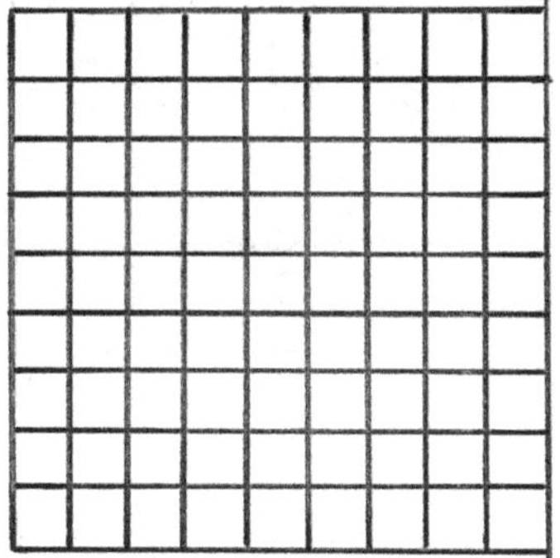

ALL THE EIGHTS

EXERCISE 3.13

Understanding that the 81 digits of the Vedic Square are each represented by a dot at the centre of each cell, locate the 6 Points of 1 upon this

DIGIT SUMS CRACK THE ATOMIC CODE ! (RUTILE)

TiO_2 (rutile) structure.

Co-ordination 6:3.

The metal ion is surrounded by six non-metal ions at the corners of an octahedron; and each non-metal ion is surrounded by three metal ions at the corners of an equilateral triangle.

Examples:- TiO_2, MnO_2, MgF_2.

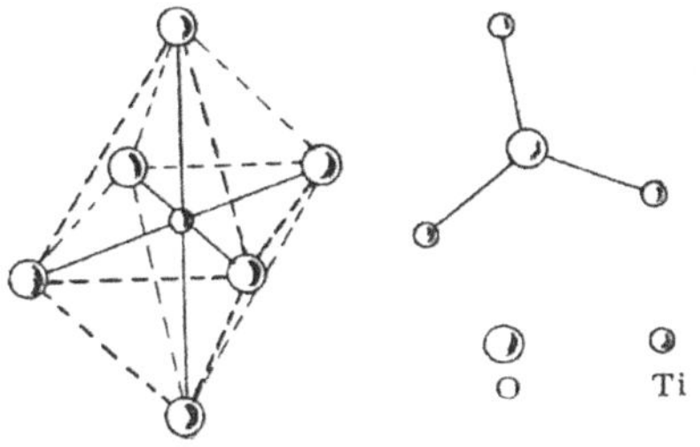

grid or matrix of 81 dots, by marking them very lightly with a small circle in pencil.

Understanding that each of the 6 digits of 1 are connected to one another in an invisible energy field, draw lines from each dot to connect with every other dot, to make the crystal-line field visible.

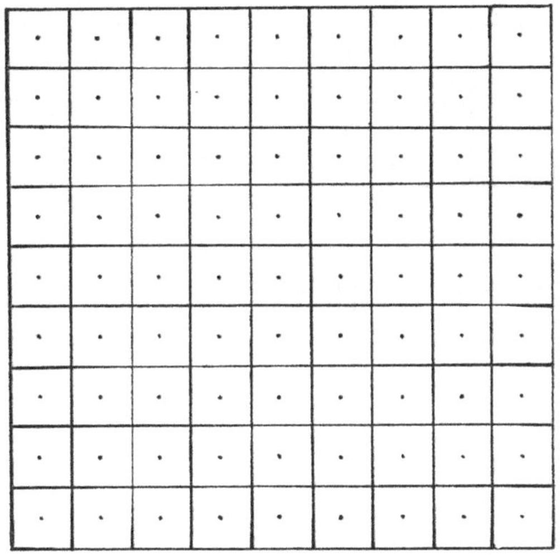

QUESTION : How would you describe the shape drawn as a 3-Dimensional form ?

EXERCISE 3.14

Understanding that All The Ones and All The Eights are similar patterns but at a 90° rotation to each other, draw both patterns on top of one another ie a super-imposition or combination for

SUPER-IMPOSING "ALL THE ONES" AND "ALL THE EIGHTS" CREATING THE ATOMIC STRUCTURE OF PLATINUM CRYSTAL!

nb: This Pattern, SuperImposed upon itSelf at 90°, makes the Front Cover Art!

a very interesting outcome, and scientific discovery! The 81 dots are provided without the frame, thus each dot is still the centre of a Cell. Use 2 colors for the 1s and 8s.

ATOMIC ART

With a simple knowledge of Digit Sums, you have just entered the world of nuclear geometry! You have just created the atomic structure of Platinum Crystal!

Shown overleaf is a rare photo of a dome or hemi or half sphere of Platinum Crystal magnified 750,000 times such that the resolution shows single atoms (by the late Erwin. H. Mueller of Penn State University). Notice the pyramid shapes of light which are the central geometric form of the macro and micro worlds. Our blood crystals are identical to

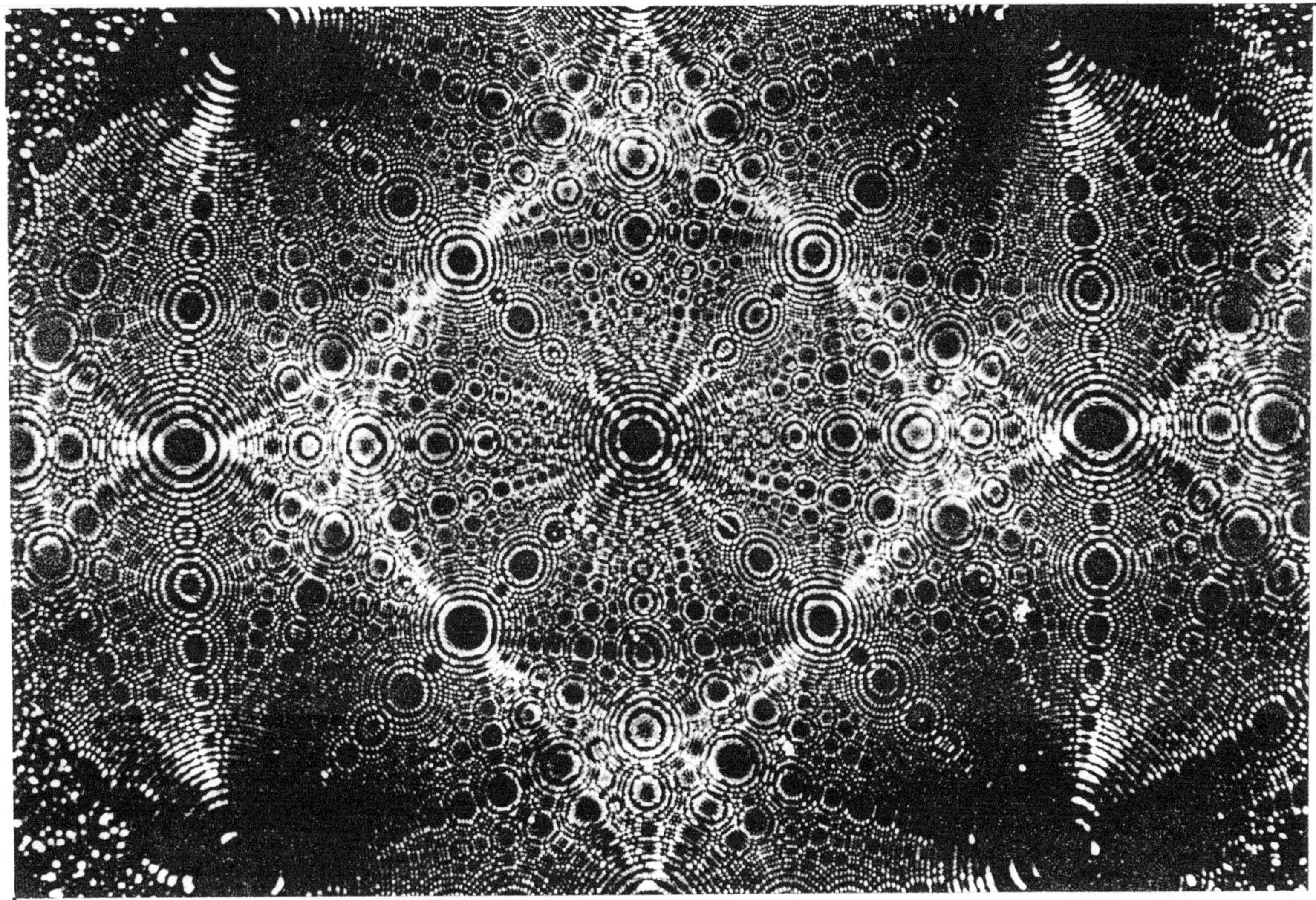

these pyramids of light as seen under Tunnelling Electron Microscopes.

If you were to look at this Platinum crystal photo under a light or against a window or actually look through the back of this page, you will see exactly every line that you just drew in Exercise 3.14 !

DIGIT SUMS

are the backdoor into Atomic Physics !

Reprinted from "THE KEYS OF ENOCH" by J. J. Hurtak, Academy of Future Science. (ATOMIC ART OF PLATINUM CRYSTAL by E. H. MUELLER).

the VEDIC SQUARE

formerly called: "The Visual Multiplication Table and it's Reduction Grid".

1	2	3	4	5	6	7	8	9
2	4	6	8	10	12	14	16	18
3	6	9						
4	8							
5								
6								
7								
8								
9								

1	2	3	4	5	6	7	8	9
2	4	6	8	1	3	5	7	9
3	6	9	3	6	9	3	6	9
4	8	3	7	2	6	1	5	9
5	1	6	2	7	3	8	4	9
6	3	9	6	3	9	6	3	9
7	5	3	1	8	6	4	2	9
8	7	6	5	4	3	2	1	9
9	9	9	9	9	9	9	9	9

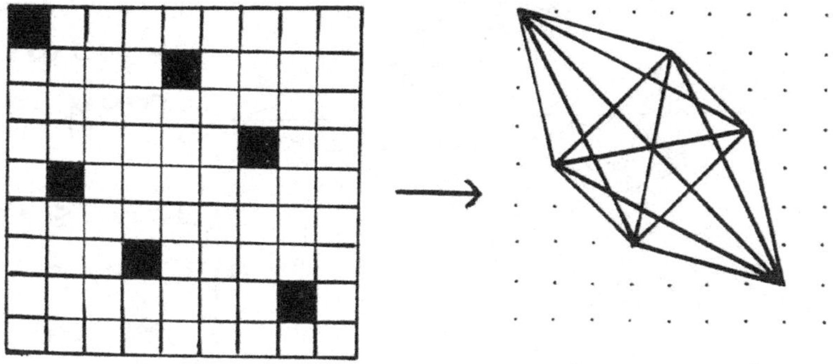

ALL THE ONES

Rutile

ALL THE EIGHTS

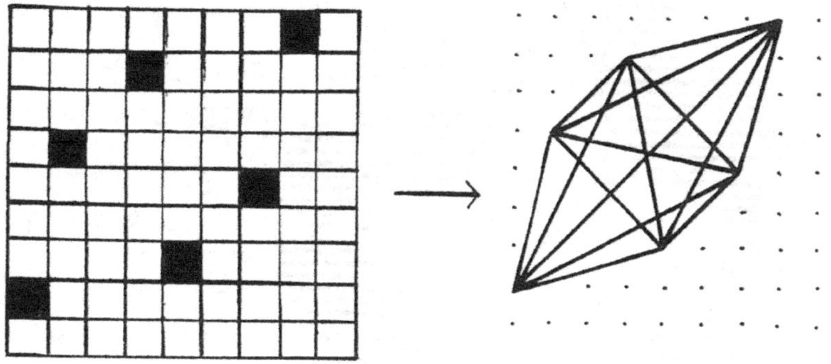

• DISCOVER HOW THE TRADITIONAL MULTIPLICATION - TABLE CREATES THE ATOMIC STRUCTURE OF:

1): RUTILE (on this page)

2): PLATINUM CRYSTAL (overleaf)

by understanding the Complements of 9! and DIGITAL SUMS!

See "THE BOOK OF MAGIC SQUARES" Volume 3. Chapter "X".

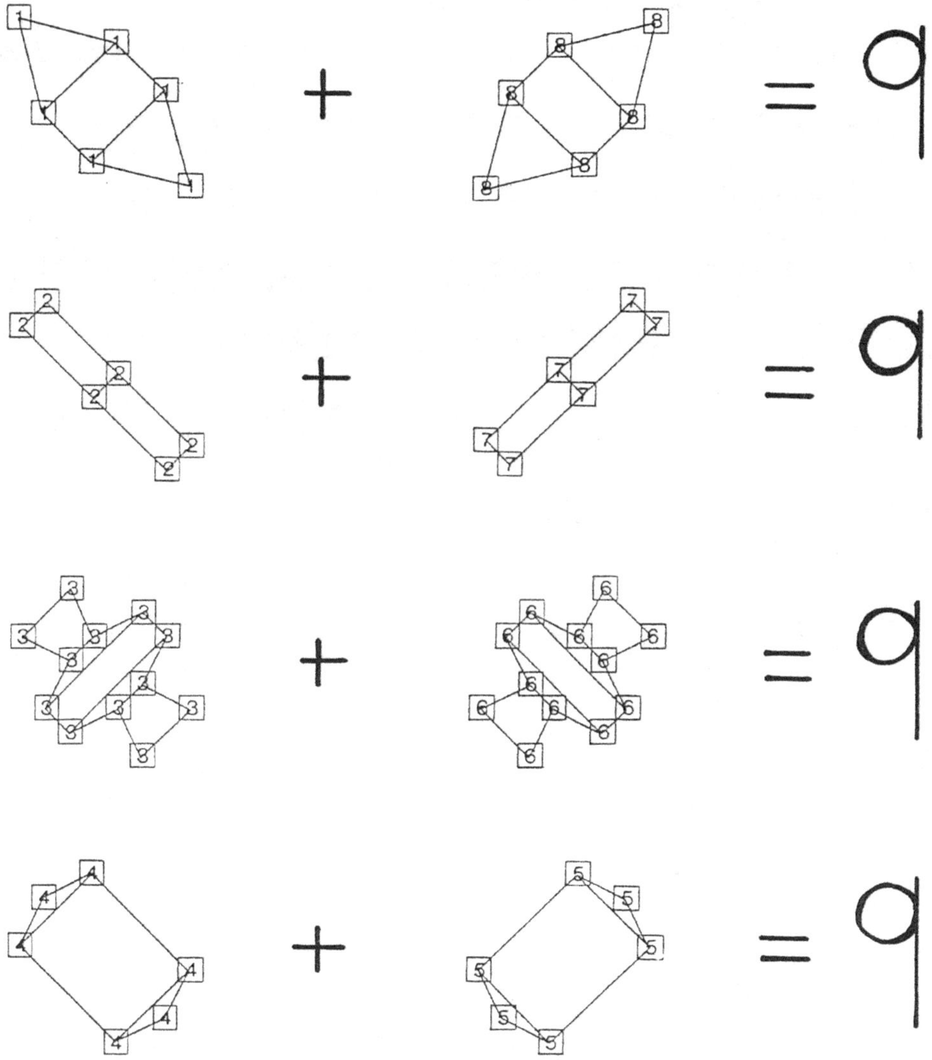

By combining the Digital Sum Pattern of All The Ones and All The Eights, the Nuclear Geometry of Platinum Crystal is created.

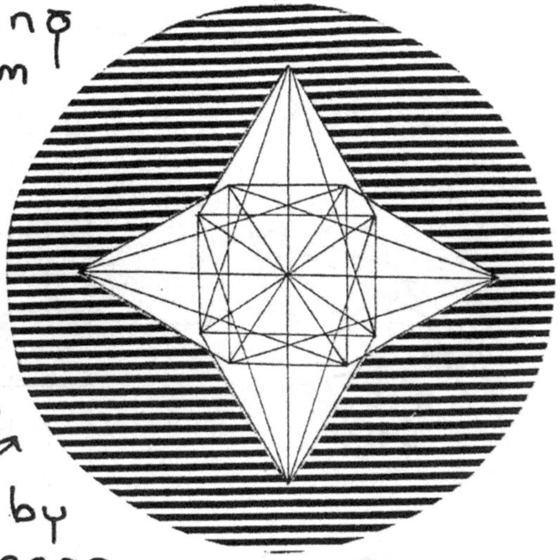

This is verified by electron microscope Atomic Art in "The Book of Magic Squares" Vol 3, thanks to the late Mueller's work.

Notice how the Complementary Patterns of Nine are identical but at a 90° rotation!

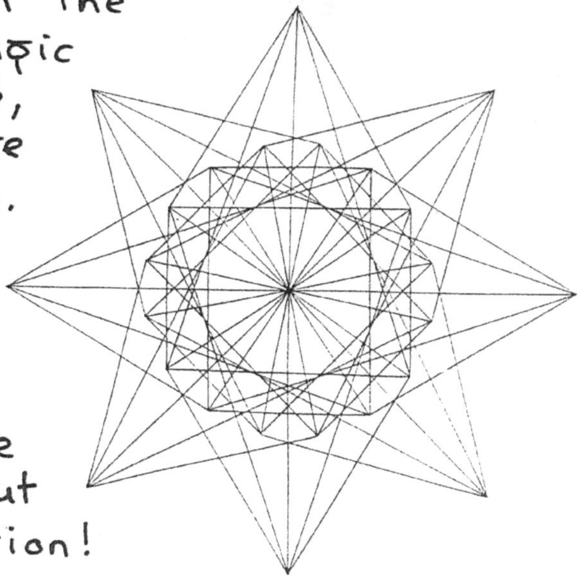

1	2	3	4	5	6	7	8	9
2	4	6	8	1	3	5	7	9
3	6	9	3	6	9	3	6	9
4	8	3	7	2	6	1	5	9
5	1	6	2	7	3	8	4	9
6	3	9	6	3	9	6	3	9
7	5	3	1	8	6	4	2	9
8	7	6	5	4	3	2	1	9
9	9	9	9	9	9	9	9	9

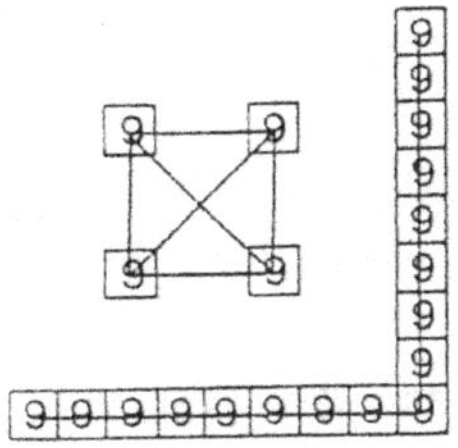

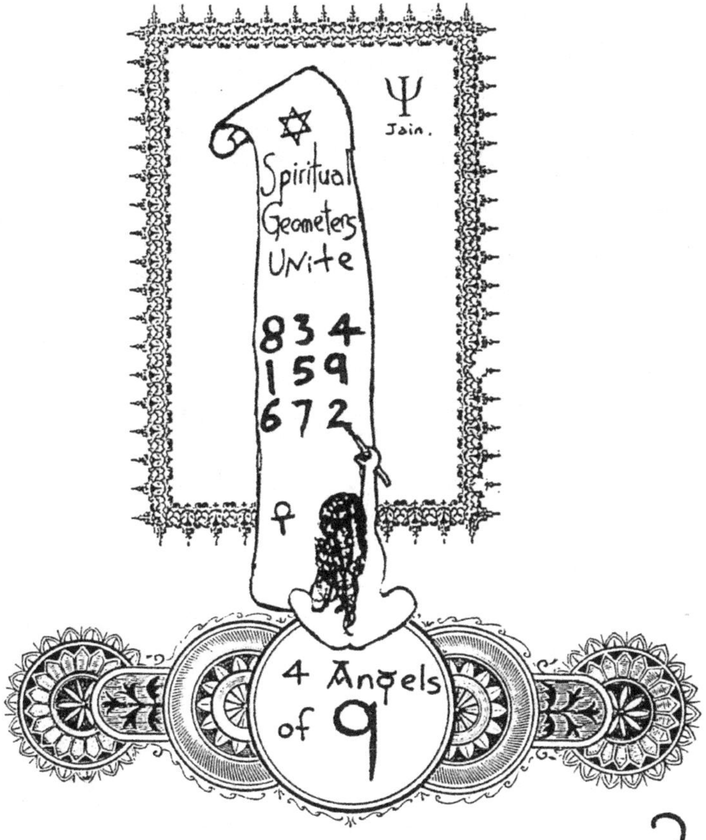

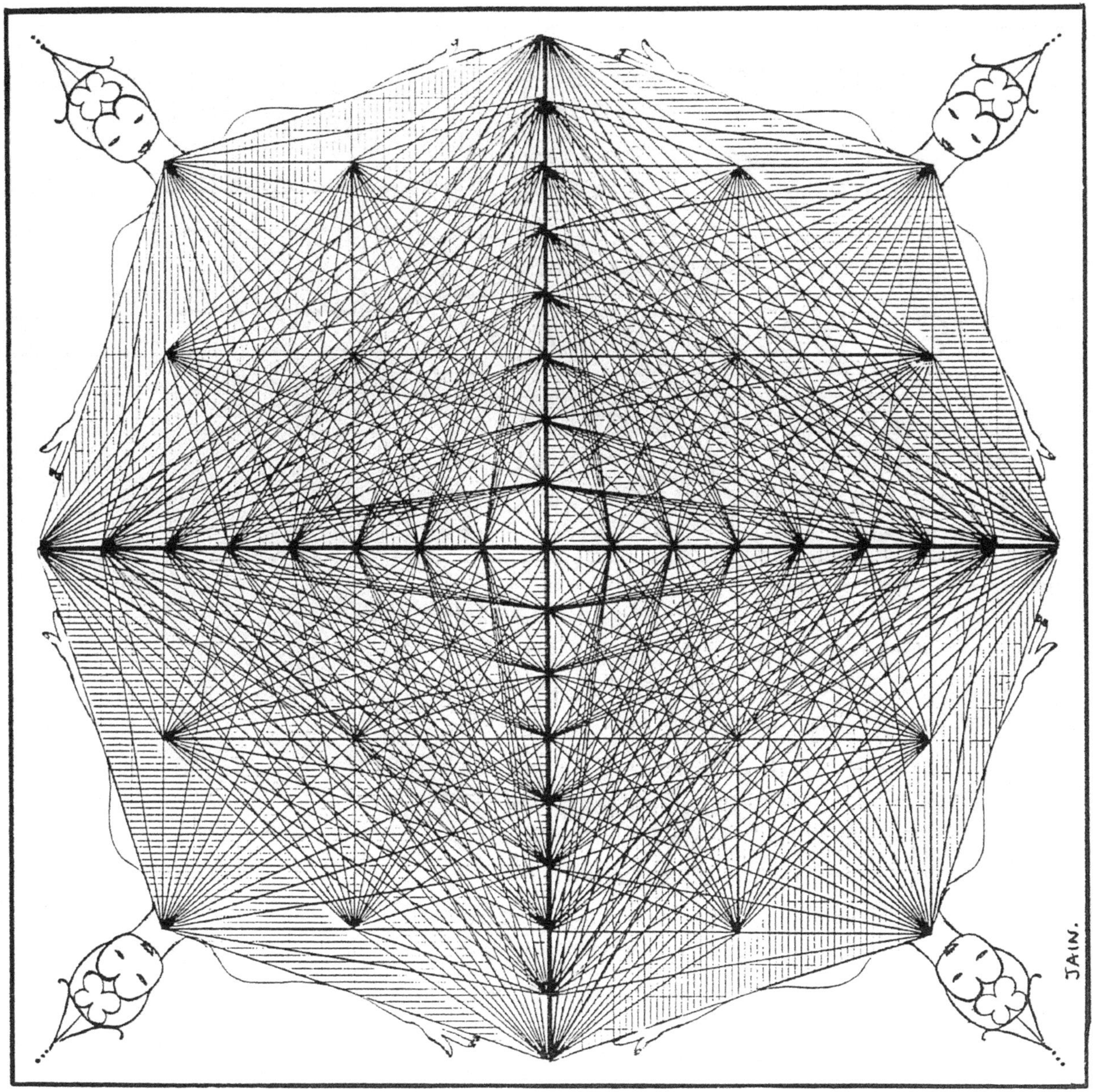

DIGIT SUM CHECK FOR DIVISIBILITY BY 3 and 6

By studying the VEDIC SQUARE shown on this page, it is apparent, that the 3 Times Table, or the 3rd Row has its Digit Sums in this Order :

3, 6, 9, 3, 6, 9, 3, 6, 9

It is a mathematical truth that:

The 3 Times Table has a Digital Sum Series Pattern of 3, 6, 9 and a Periodicity (P) = 3

ANY NUMBER DIVISIBLE BY 3 OR 6 MUST HAVE ITS DIGIT SUMS ADDING TO 3, 6 OR 9.

eg Is 3702 Divisible by 3?

3702 = 3+7+0+2 = 12
and 12 = 1+2 = 3
∴ Yes, it is divisible by 3.

1	2	3	4	5	6	7	8	9
2	4	6	8	1	3	5	7	9
3	6	9	3	6	9	3	6	9
4	8	3	7	2	6	1	5	9
5	1	6	2	7	3	8	4	9
6	3	9	6	3	9	6	3	9
7	5	3	1	8	6	4	2	9
8	7	6	5	4	3	2	1	9
9	9	9	9	9	9	9	9	9

eg Is 1175 Divisible by 21?

In this case, we first look at the Factors of 21, ie the numbers, when × together, give 21. They are 3 and 7.
We now test for divisibility by 3 firstly, then 7 secondly.
1175 = 1+1+7+5 = 14 = 5
Since 5 is not ÷ by 3, 6 or 9 then divisibility is not possible.
We need not check for divisibility by 7. ∴ NO.

NB : THE 6 TIMES TABLES HAS A DIGITAL SUM SERIES PATTERN OF : 6, 3, 9 AND A PERIODICITY OF (P) = 3

EXERCISE 3.15

Check that the following nos. are divisible by the no. shown:

a. 1998 ÷ 3
b. 7404 ÷ 6
c. 74068 ÷ 3
d. 14806 ÷ 6
e. 18510 ÷ 15
f. 259244 ÷ 21
g. 44424 ÷ 18
h. 222212 ÷ 18

There is a Western Law for the DIVISIBILITY OF 4 :

IF 4 DIVIDES INTO THE TWO LAST FIGURES OF A NUMBER THEN 4 DIVIDES INTO THE WHOLE NUMBER.

- eg Is 1848 Divisible by 4 ?

The Best Way to Divide by 4 is by halving twice.
We see that the last two digits of 1848 = 48
We halve it once, 48 ÷ 2 = 24
We halve 24 = 12. Since 12 is divisible by 4, then 1848 is ∴ divisible by 4.

- nb : Nos. ending in 00 are ÷ 4.

EXERCISE 3.16

Using both Western + Eastern methods check Divisibility by 4.

a. 4984
b. 22712
c. 26170
d. 3156
e. 26664
f 22528

DIVISIBILITY BY 4

EMPLOYING THE VEDIC SUTRA: "THE ULTIMATE AND TWICE THE PENULTIMATE".

There is also an Eastern Law for the DIVISIBILITY of 4 : It involves one of the Vedic Sutras called :

SOPANTYADVAYAM'ANTYAM

which translates from the Sanskrit language as :

THE ULTIMATE AND TWICE THE PENULTIMATE.

The last digit is called the 'Ultimate' and the digit before the last one is called the 'Penultimate'.
eg Is 34552 divisible by 4 ?
To apply this Sutra we need to double or ×2 the second last digit which is 5 thus 5 × 2 = 10, and add to this the last digit, which is 2, thus 10 + 2 = 12. Since 12 is divisible by 4 then the whole number of 34,552 is divisible by 4.

The ROWS of the VEDIC SQUARE'S D.S.S.P PLOTTED upon the 9-POINT CIRCLE

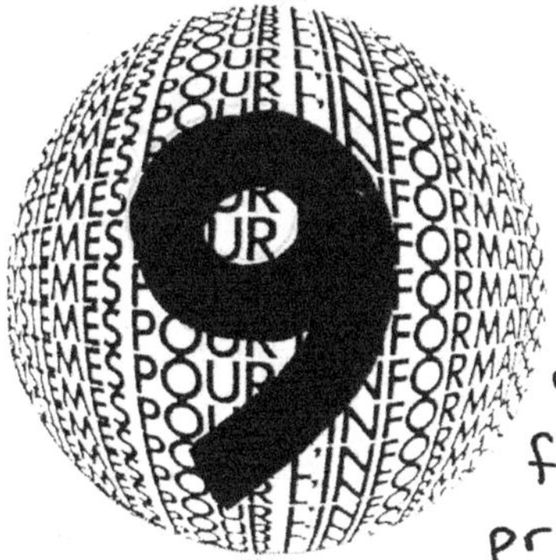

We can use the 9-Point Circle to look for hidden though striking patterns within the Vedic Square.

The following 4 preliminary examples will assist you in creating 4 superb diagrams giving valuable insight into the nature of the mysterious number 9

eg 1: Regarding Row 1 of the Vedic Square, plot its D.S.S.P (Digital Sum Series Pattern) of:

1, 2, 3, 4, 5, 6, 7, 8, 9 upon the 9-Point Circle. This is achieved by drawing a long continuous line, according to the order of the numbers from 1 → 2 → 3 → 4 → 5 → 6 → 7 → 8 → 9.

To know where the Pattern starts from, place a bar above the no. 1 ($\bar{1}$)

To know where the Pattern ends, place a bar below the $\underline{9}$:

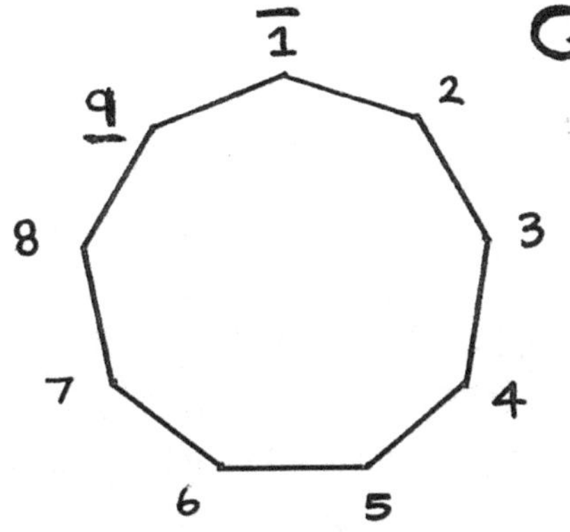

eg 2: Regarding Row 8 of the Vedic Square, plot its D.S.S.P of:

8, 7, 6, 5, 4, 3, 2, 1, 9 upon the 9-Point Circle:

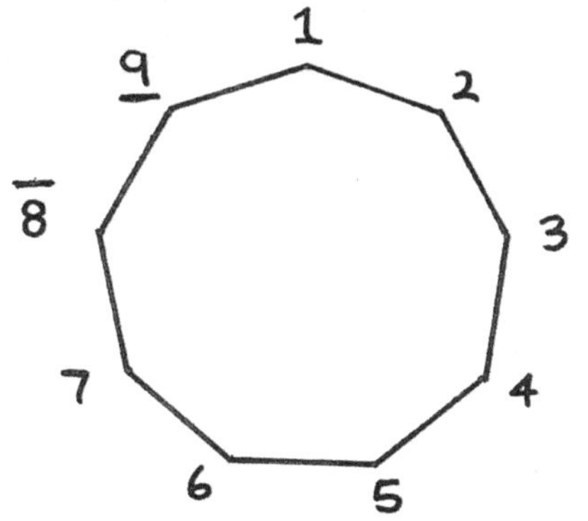

You can conclude that the D.S.S.P for 1 and 8 of the Vedic Square produce similar 9-sided ENNEAGONS (from the Latin word "NONUS" = ninth).

eg 3: Likewise, the Pattern for Row 3, having a D.S.S.P of 3, 6, 9, looks like this:

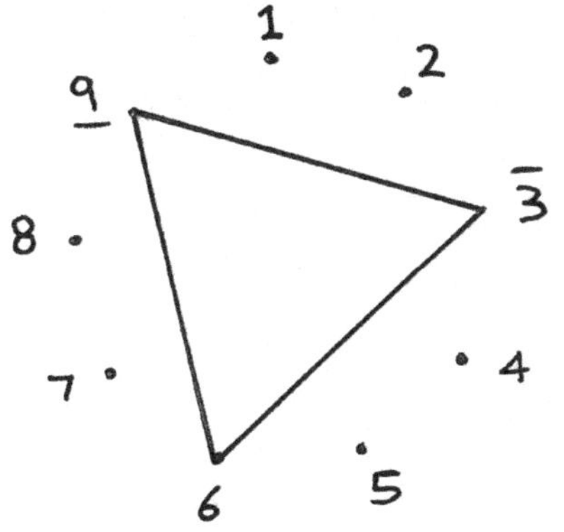

eg 4: The Pattern for Row 6 has a D.S.S.P of 6, 3, 9 and looks like this:

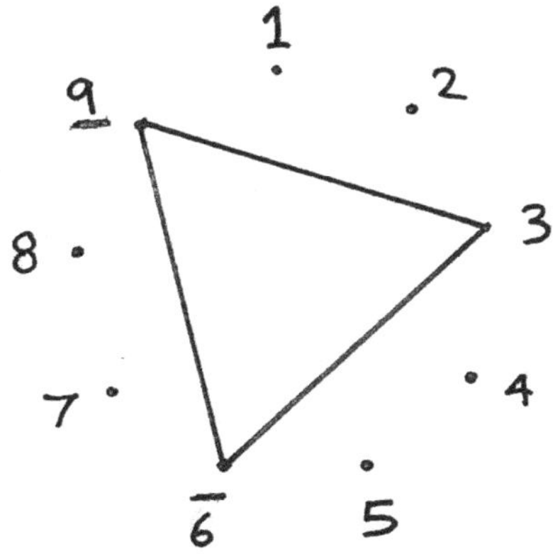

The D.S.S.P for ROWS 3+6 of the VEDIC SQUARE on the 9-POINT CIRCLE

You will have noticed that the pattern for 3 and 6 are the same.

Notice that the Patterns for 1 + 8 = 9
and 3 + 6 = 9

We could also conclude that the Patterns for the Complements of 9 move in opposing or reverse directions.

EXTRA NOTES:

• Another view of what a Digit Sum is to see it as a remainder after being divided by 9
eg 226 ÷ 9 = 25 remainder 1.
∴ its Digit Sum is 1.
Normally we just add the digits and say:
226 = 2+2+6 = 10 = 1+0 = 1.
Because it took two steps i.e: 226 → 10 → 1, it has a PERSISTENCE NUMBER (P.N.) = 2
eg: 324 = 3+2+4 = 9 ∴ PN = 1
eg: 289 → 19 → 10 → 1 ∴ PN = 3.

The D.S.S.P for ROWS 2+7 of the VEDIC SQUARE on the 9-POINT CIRCLE

EXERCISE 3.17

Upon the 9-Point Circle provided, connect in the given order, the nos. of Row 2 of the Vedic Square whose D.S.S.P = 2, 4, 6, 8, 1, 3, 5, 7, 9

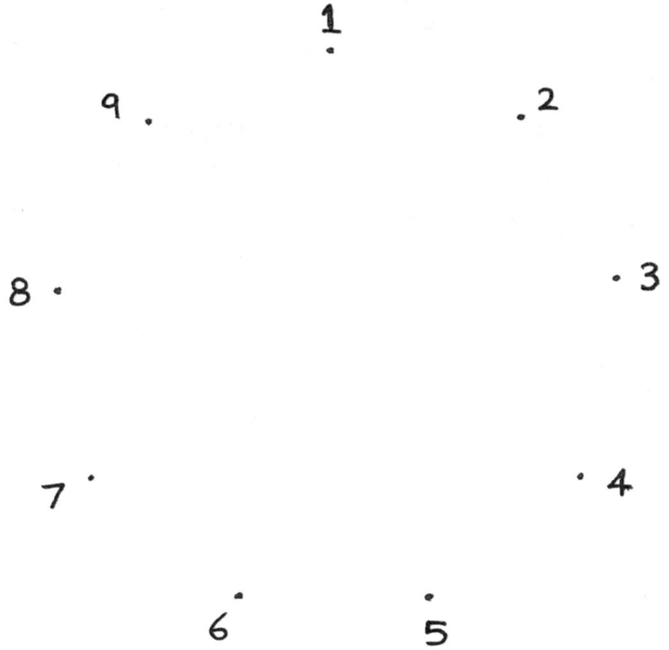

EXERCISE 3.18

Upon the 9 - Point Circle provided, connect in the given order, the nos. of Row 7 of the Vedic Square whose D.S.S.P. = 7, 5, 3, 1, 8, 6, 4, 2, 9

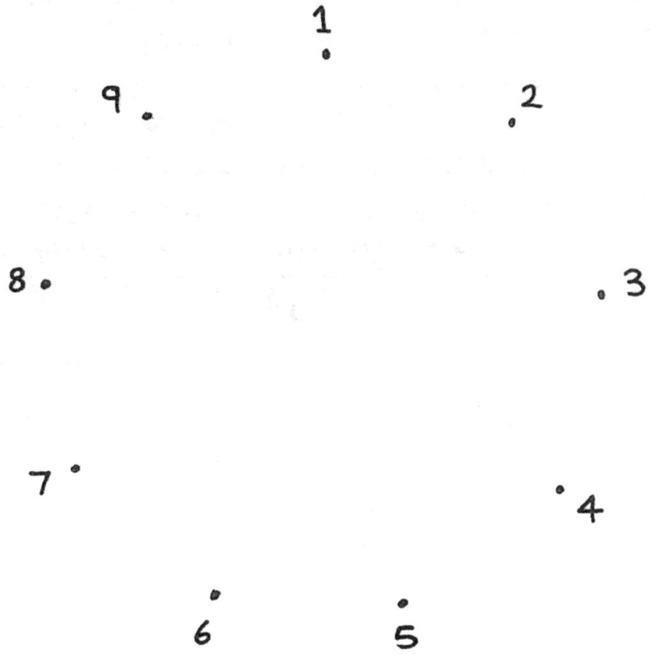

Beauty in Mathematics

"BEAUTY IS A WORD OF GOD"

QUESTION: What is the obvious relationship between the 9-Point Circle Patterns for Rows 2 and 7 of the Vedic Square.

The D.S.S.P. for ROWS 4 + 5 of the VEDIC SQUARE on the 9-POINT CIRCLE

EXERCISE 3.19

Upon the 9 - Point Circle provided, connect in the given order, the nos. of Row 4 of the Vedic Square whose D.S.S.P = 4, 8, 3, 7, 2, 6, 1, 5, 9

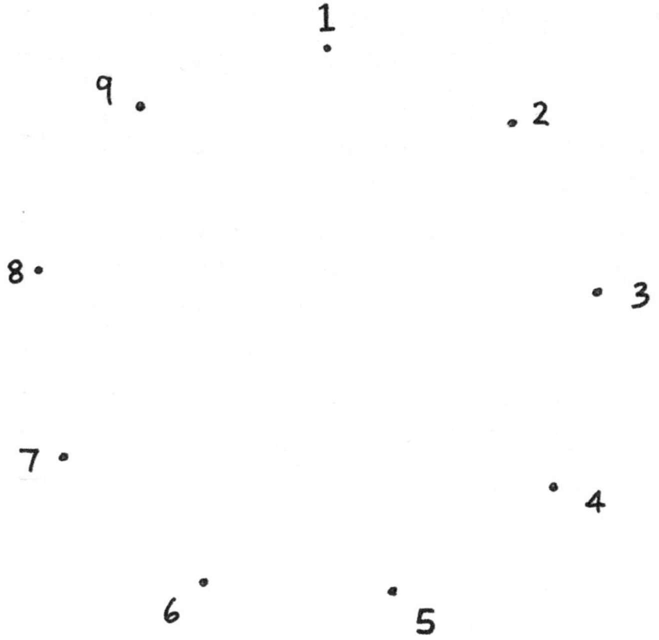

QUESTION 1 :
What is the technical name for this shape constructed ?

EXERCISE 3.20

Upon the 9-Point Circle provided, connect in the given order, the nos. of Row 5 of the Vedic Square whose D.S.S.P = 5, 1, 6, 2, 7, 3, 8, 4, 9

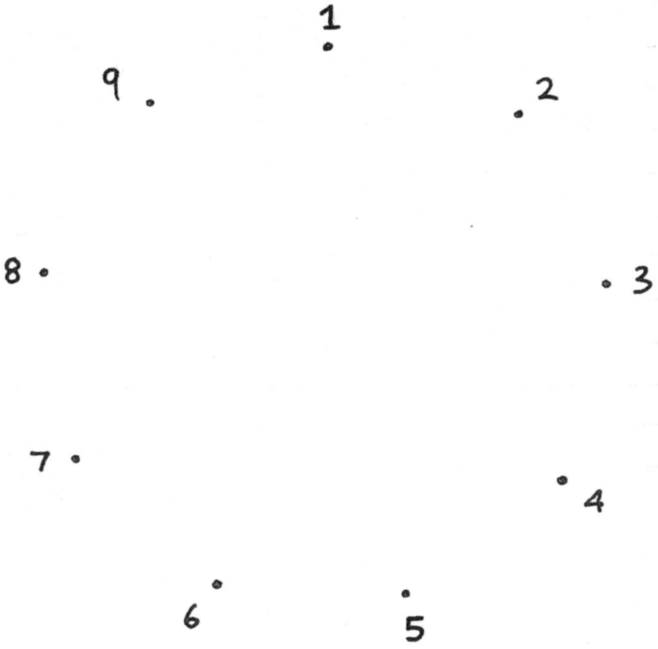

QUESTION 2: What is the obvious relationship between the 9- Point Circle Patterns for Rows 4 and 5 of the Vedic Square ?

PROBING DEEPER INTO EXISTING PATTERNS

We have identified that the Digital Sum Series Pattern for Row 4 of the Vedic Square = 4, 8, 3, 7, 2, 6, 1, 5, 9.

At first sight, there appears to be no obvious recognition of order.

If we examine or probe deeper into the 4 Times Table, from 4 x 8 ⟶ 4 x 16, we can detect another pattern operating hitherto unknown:

Study the Chart on the right. By arranging the column of the 9 Digit Sums into two distinct columns, you will observe a surprising pattern that alternates two columns of consecutively ordered nos. (ie: in natural counting order or sequence 1→2→3).

ORDER Within APPARENT DISORDER!

4 TIMES TABLE	D. SUM	D.S. ARRANGED INTO 2 COLUMNS	
4 x 1 = 4	4		
4 x 2 = 8	8		
4 x 3 = 12	3		
4 x 4 = 16	7		
4 x 5 = 20	2		
4 x 6 = 24	6		
4 x 7 = 28	1		
4 x 8 = 32	5	5	
4 x 9 = 36	9		9
4 x 10 = 40	4	4	
4 x 11 = 44	8		8
4 x 12 = 48	3	3	
4 x 13 = 52	7		7
4 x 14 = 56	2	2	
4 x 15 = 60	6		6
4 x 16 = 64	1	1	

(D. Sum = Digit Sum)

CHAPTER 4

EKĀDHIKA

and

$$\frac{1}{19}$$

DIGIT SUMS

CHAPTER 4

DIVISIBILITY, OSCULATORS, EKĀDHIKĀ and $\frac{1}{19}$

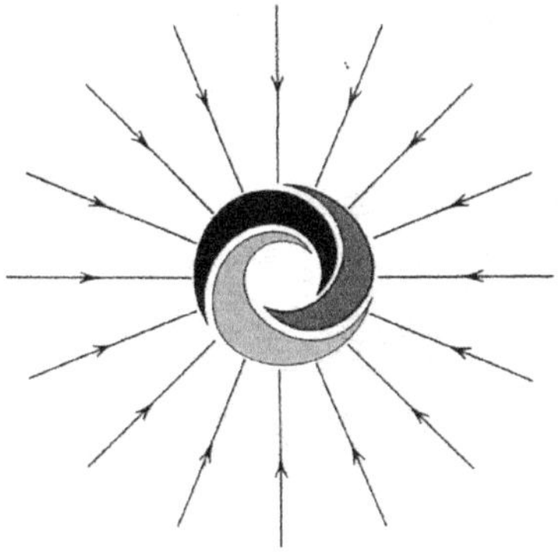

The process of knowing answers to questions like: "Is 50,594 Divisible by 41?" comes under a fascinating subject called "Simple Osculators". It will be dealt with in full, not in this chapter, but in a forthcoming booklet just on the many diverse methods of Vedic Division. It is merely referenced here, in Digit Sums, as it examines every digit of that number being tested eg 50,594. Digital examination extracts good information from the many parts or digits to know more of the whole; an intelligent movement from the small fractal digital Branch to the tree of Unity or Wholeness.

This subject of Divisibility (or Osculation, from the Sanskrit word "Vestana" and Latin word for "kiss") must first start with a knowledge of "EKADHIKA". For more in-depth information on 'Osculators', refer to the Masters book on "Vedic Mathematics". We have already tested for Divisibility by 11 in Exercises 2.9 and 2.10 which looked at the relationship of the oddly and evenly placed digits, a simple and ingenious method.

WHAT IS THE EKĀDHIKĀ ?

The Ekadhika is "The number 'ONE MORE' than the One Before, when the number ends in a Nine"

To derive this, we must have a "9" in the unit's place:

eg the Number 39 has the '9' in the correct place, and we say "One More than 3" is 4

Thus 4 is the Ekadhika!

eg What is the Ekadhika of 23 ?
To get a 9 in the unit's place, we multiply 23 by 3 = 69. We say, "One More than 6 is 7"
∴ The Ekadhika = 7.

EXERCISE 4.1

What is the Ekadhika of the following numbers ?

a. 19
b. 69
c. 139
d. 9
e. 7
f. 13
g. 17
h. 11
i. 27
j. 3

Now that you understand this most important Sutra: "BY ONE MORE THAN THE ONE BEFORE" or (Ekādhika Purvena in the Sanskrit language), you will discover its many

"EKĀDHIKA PŪRVENA"

BY ONE MORE THAN THE ONE BEFORE

CREATION of the NATURAL COUNTING NUMBERS.

applications. Here are 2 examples :

1) The Creation of the Natural Counting Numbers
2) Converting Fractions to Decimals. eg $\frac{1}{19}$ = $\cdot\dot{0}5263157894736842\dot{1}$

1) Vedic Maths or Arithmetic has Unity at its Source. The Number 1 is the symbol of wholeness as seen ubiquitously in Nature and the Cosmos. From this springs forth the family of whole numbers by applying this primal and universal (uni = 1 and verse = spin) Sutra "By One more Than The One Before" which generates 2, and then One More Than 2 is 3, then 3 leads to 4, to 5, 6, 7, 8 and 9 etc.

1,2,3,4,5,6,7,8,9

2) The SUTRA: "EKADHIKENA PURVENA" also assists in solving difficult divisions like: $1 \div 19$, which is really the decimalizing of a vulgar fraction ($\frac{1}{19}$).

By merely observing the shape of the fraction $\frac{1}{19}$ the Vedic Mathematician can adeptly apply a mental and One-line calculation by firstly "seeing the predicted answer" as a beautiful numerical landscape that he/she will paint, saying to their Self

[What follows is a list of 10 salient features A→J, which in the end, shows brilliance in the field of Digit Sums/ Complements of 9 application]

A) By knowing "BY ONE MORE" than the 1 in $\underline{1}9$, I have quickly determined, from the very onset, that 2 is my specific Ekadhika, or Multiplier, in this case.

B) I know confidently how

Chapter 4.

A Knowledge of EKADHIKA can reduce the conversion of $\frac{1}{19}$, of the Western Way, which is a shocking mess of 35 descending lines of working out, to a mere 1 VEDIC Line!

Exposition on $\frac{1}{19}$

many digits there will be in the answer: Denominator (D) or the bottom digits minus the Numerator (N) or top digit: $D-N = 19-1 = 18$ digits. (Wow. I never got taught that at school, but at least I know now!).

C) I know that the last digit of the 18 digit answer will be the Numerator 1 in $\frac{1}{19}$ because I have the 9 in the correct place as the ultimate digit.

D) I know that the answer will recur every 18 digits since Denominator D or 19 is a prime number which means that it is indivisible by any other number. In mathematical jargon or symbols, this forseen digital circulation will be indicated by 2 dots, one at the beginning and one at the end, with both dots placed above the digits. A necklace of 18 digits going to Infinity.

E) I know that when I get to D−N or 19−1 = 18 as an answer digit I am half way or 9 digits along, whilst moving from Right ⟵ to Left, that is working in reverse, at which point I can choose to stop multiplying by 2 and merely take the Complements of 9 to complete the answer. Here are the first 9 digits :

= • ________ / 9,4 7,3,6 8 4 2 1

derived by a) Last Digit = 1.
b) 1 × 2 = 2 c) 2 × 2 = 4, d) 4 × 2 = 8.
e) 8 × 2 = 16. Put down the 6, and carry 1. f) 2 × 6 = 12 + 1 = 13. Put down the 3 and carry 1, g) 3 × 2 = 6 + 1 = 7. h) 2 × 7 = 14. Put down 4 and carry 1. i) 4 × 2 = 8 + 1 = 9.

F) = • ________ / 9,4 7,3,6 8 4 2 1
0 5 2 6 3 1 5 7 8

I can place the last 9 digits

7 DAY WORKSHOP

VEDIC MATHEMATICS

"TEACHER TRAINING"

UNDER THE AUSPICES OF

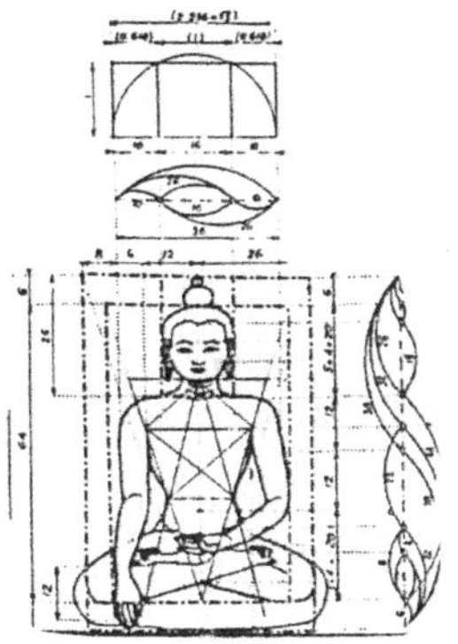

under the first 9 digits, by taking The Complements of 9. There are now 9 columns adding to 9 eg 1+8 = 9, 2+7 = 9, 4+5 = 9 etc.

G) If I didnt choose 'Complements of 9' I could carry on by × 2 to get the 18 digits, although its harder work :

= • 0 5,2 6 3,1,5,7,8 / ________ 1

derived from: j) 9 × 2 = 18. Put down the 8, carry 1. k) 8 × 2 = 16 + 1 = 17. Put down the 7, carry 1. l) 7 × 2 = 14 + 1 = 15. Put down the 5, carry 1. m) 5 × 2 = 10 + 1 = 11. Put down the 1, carry 1.
n) 1 × 2 = 2 + 1 = 3 o) 3 × 2 = 6
p) 6 × 2 = 12. Put down the 2, carry 1. q) 2 × 2 = 4 + 1 = 5
r) 5 × 2 = 10. Put down the 0, carry 1.
We notice after these 18 steps that recursion begins, so we put up the recursion marks and say :
$\frac{1}{19} = \cdot\dot{0}52631578/94736842\dot{1}$.

H) Even after this answer, I still know a lot more:
I know that if I chose not to multiply by 2 and start from R ← L, from last digit 1 to first digit **Zero**, I could have as easily DIVIDED by 2 and started from L → R, (from Left → to Right):

$\frac{1}{19} = {}^{\bullet}{}_1 0\ 5\ {}_1 2\ 6\ 3\ {}_1 1\ {}_1 5\ {}_1 7\ {}_1 8 /$

by mentally saying:
a) 1 ÷ 2 = ·0 with a rem 1. I prefix this remainder 1 infront of the Zero, like ₁0, which really is "10". b) 10 ÷ 2 = 5.
c) 5 ÷ 2 = 2 rem 1. Prefix the rem 1 infront of the 2 to make 12
d) 12 ÷ 2 = 6 e) 6 ÷ 2 = 3
f) 3 ÷ 2 = 1 rem 1 = 11
g) 11 ÷ 2 = 5 rem 1 = 15
h) 15 ÷ 2 = 7 rem 1 = 17
i) 17 ÷ 2 = 8 rem 1 = 18 ← AHA!
I know that when I reach this magical number of D-N or 19 − 1 = 18, an AHA or

WHAT
FINER
EXAMPLE
THAN
THIS
IS
THERE
THAT
CLEARLY
DEMONSTRATES
THE
MAIN
VEDIC
PRINCIPLE
OF
UNIVERSAL
SHAPE
OR
PATTERN
RECOGNITION

sigh occurs indicating a doorway or Gate has opened at the ½ way mark which tells me to stop the division by 2 and merely reel off the remaining 9 digit answers mentally as the Complements of 9.

I) I also know the internal synchronicities: that if I line up these 2 rows of Complements of 9, with their carry-overs and employ a unique type of digital summing, I have:

9 ₁4 7 ₁3 ₁6 8 4 2 1
₁0 5 ₁2 6 3 ₁1 ₁5 ₁7 ₁8

which is really 9 columns that all add up to 19!
eg from L → R 9 + 10 = 19
14 + 5 = 19, 7 + 12 = 19 etc.

All this information, because we started with the correct application, The Truth of The Ekadhika, a mere "ONE MORE".

$\frac{1}{19} = \cdot\dot{0}52631578/94736842\dot{1}$

J) The Tenth Insight:
Another way of achieving insight into the internal/underlying harmony of digital arrangements is to plot all the Digits on a 9-Point Circle (see Exercise 2.2).

EXERCISE 4.2

Plot the circulating 18 Digits of $\frac{1}{19}$, starting from the left-most digit zero.
nb: The Zero can be plotted as a '9'. (Sometimes a 10-Point Circle can be used to accommodate The Zero).
<u>QUESTION</u>: What type of symmetry do you notice?

Connecting these 18 Digits in their sequential order, transforming Mathematics (Left-Brain, Logical, Masculine) into ART (Right Brain, Feminine Intuitive, Color, Hologram) allows WHOLE-BRAIN LEARNING and Numerical Discovery.

A PYTHAGOREAN MYSTERY REVEALED!

CONVERTING NUMBER PATTERNS INTO SOUND!

"HEAR" THE MATHS

The 9-Point Circle or **ENNEAGRAM** has a practical application. 2,500 years ago Pythagoras created Healing Music by allocating a specific musical note for a specific pitch to each of the 9 digits on the Circle, and violà, any pattern can be heard.

This means that all the Digital Sum Series Patterns that you discovered in Chapter 3, not only can be 'seen' by Digital Sums, but by the allocation of musical notes, you can "hear" the mathematics.

This revelation has unlimited scientific potential.

Music is a mathematically based system relating to the doubling or halving of Numbers as observed in Pascal's Triangle: =1, 2, 4, 8, 16, 32, 64 etc, being the Law of the Octave.

CHAPTER 5

ADDITION AND SUBTRACTION

CHAPTER 5

an Introduction to ADDITION and SUBTRACTION

AMBIDEXTERITY

This chapter will focus on addition, subtraction and an introduction to a sub-sutra "Dhvajanka" or "On The Flag".

Of most importance is the ability to achieve the same result by approaching from different angles or views. This is the ability to be AMBIDEXTROUS, to solve mathematical problems, like subtraction and multiplication by moving from the Right and proceeding to the Left, or conversely, from Left to Right. This MENTAL AMBIDEXTERITY applies to a good dancer who can perform simple choreographed routines from one direction and then in reverse.

Similarly, in Yoga (means 'Union') true balance is achieved when breathing and body stretches are performed diligently from Left to Right and Right to Left. What we do to one side, to maintain EQUI-STASIS, we must do to the other.

A good mathematician is not afraid to learn new numerical acrobatics. Just to master the diverse use of Digital Sums will enhance your Education and you will begin to appreciate why the whole global mathematics curriculum must begin with a thorough yet simple application of DIGIT SUMS.

THE DIGIT SUM OF THE SUM IS THE SUM OF THE DIGIT SUMS

DIGIT SUM CHECKING

There is an important Vedic Sub-Sutra called :

GUNITA SAMUCCAYAH
SAMUCCAYAGUNITAH

which is used chiefly in Digit Sum Checking. translated, it means :

THE PRODUCT OF THE SUM IS THE SUM OF THE PRODUCTS

Since "Product" refers to "Multiplication", this Sub-Sutra has an adapted form to apply for Digit Sum Checking of Addition and Subtraction. the same Sub-Sutra can read as :

THE DIGIT SUM OF THE SUM IS THE SUM OF THE DIGIT SUMS

Here is a simple example.

Is

$$\begin{array}{r} 123 \; + \\ 456 \\ \hline 579 \end{array}$$

correct ?

We work out the Digit Sum of the 3 lines and thereby create a simpler sum.

- Digit Sum of 123 = 1+2+3 = 6
- Digit Sum of 456 = 4+5+6 = 6 (by casting out the 4+5 or 9)
- Digit Sum of 579 = 5+7 (by casting out the 9) = 12 = 3.

We tabulate this new information like this :

$$\begin{array}{rr} 123 \; + & 6 \; + \\ 456 & 6 \\ \hline 579 & 3 \end{array}$$

Since 6+6 = 12 = 1+2 = 3 and the Digit Sum of the Answer is 3, we then have a confirmed Yes.

EXERCISE 5.1

Add these numbers and check all your answers using Digit Sums, following the format just shown:

a. 23 + 45

b. 67 + 73

c. 145 + 623

d. 246 + 378

e. 12345 + 67890

A CURIOSITY

Sometimes, when you have unawarely made an error in your addition, and you have checked your Digit Sums as correct, there still may be an error!

Here is an unusual example:

"The Product of the SAMUCCAYA (or DIGIT SUM) is Equal to the SAMUCCAYA of the Product"

"Sam-uccaya" means:

- Sum
- Aggregation
- Accumulation
- Collection
- Multitude
- Totality
- Cross-Checking
- Balance
- Factor
- Variables
- Constants
- Filling in the Gaps
- Reflection
- Symmetry

eg 479 +	⟶	D. Sum = 2 +
221	⟶	D. Sum = 5
790	⟶	D. Sum = 7

According to this Sutra, the addition is confirmed correct, but if you add the numbers again:
$479 + 221 \neq 790 = 700$!
So where is the error?
("$\neq$" means "does not equal").

Since we only have 1 ⟶ 9 + Zero = 10 digits, there is an obvious 1 in 10 chance of incurring this error. In the event of an error we can thus conclude that the Digit Sum check works 90% of the time, and to just be aware of this fact.

As outlined in the central column, one Sutra can have a vast gamut of meanings for states of Mind.

There are easier ways to add up strings of numbers. There is a Sutra called :

BY THE COMPLETION OR NON-COMPLETION

which invites you to join numbers that add up to Decimal Bases like :
10, 100, 1,000 etc
This approach to Wholeness
- reduces error
- increases swiftness of mind or mental agility
- encourages you think in terms of Number Pairs or Complements.

eg: Add these 4 numbers :
18 + 1 + 19 + 2.
If you went 18 + 1 = 19 + another 19 = 38 + 2, you would get 40 but not

CHAPTER 5

SUTRA :

"BY THE COMPLETION OR NON-COMPLETION"

a.k.a.

PŪRANĀPŪRANĀ-BHYĀM

as quick as if you joined the pairs :

18 + 1 + 19 + 2

and say 20 + 20 = 40 !

EXERCISE 5.2

Try these examples :

a. 38 + 17 + 2 + 3
b. 95 + 12 + 5
c. 8 + 33 + 12 + 7 + 9
d. 36 + 31 + 64 + 9

e.
```
 35
 83
 75
 27
----
```

f.
```
 7938
 1544
 3162
 9576
------
```

g.
```
 3933
 2402
 7143
 4022
 8216
------
```

The Art of SUBTRACTION is introduced at this stage for 3 reasons :

1) We want to practise a new technique opposite to the current method, and that is to subtract from the left-hand side and move along to the right. This AMBIDEXTERITY will be useful in the next booklet (Book 2) on Multiplication; knowing how to multiply from L ⟶ R and L ⟵ R helps build the mental muscle.

2) We will check, that our answers are correct, by using DIGIT SUMS

3) Subtraction introduces another Sub-Sutra called "On-The-Flag" which brilliantly solves long division, simply and in one-line, and will be dealt with in full in Book 3 on 'Division' : eg : if you were to divide 1,234,689 by 4,567, you

"ON THE FLĀG"

(DVAJANKA)

means :

- Remembering without effort
- Part of Intermediate Working
- Temporary

merely raise or hoist the last digits of 4,567 and write it as 4^{567}. This means that you merely divide by 4 and not 4,567 !

Here are 2 examples to subtract the new way :

1)
$$\begin{array}{r} 7\;3\;5\;2 \\ -\;2\;7\;1\;8 \\ \hline 5 \end{array}$$

- Begin from the left and say 7−2 = 5
- Before we put down "5" we must look ahead into the next column
- Since the top digit "3" is less than the bottom digit "7" (ie 3 < 7) we must reduce our original 5 in the answer by 1 and put down 4 :

$$\begin{array}{r} 7\;{}^{1}3\;5\;2 \\ -\;2\;7\;1\;8 \\ \hline 4 \end{array}$$

Also, we must carry 1 over to the 3 to make it 13.

We all learnt this at an early age, the raising of the "little One (1)" infront of a digit to cont...

give it an extra "10". This process of making the "3" to be "13" (Temporarily) is an application of yet another Sub-Sutra called "ON THE FLAG" used ingeniously in Long Division.

```
7'3 5 2 -
2 7 1 8
-------
4 6
```

• We now say 13 − 7 = 6. Put down the 6.

• Say 5 − 1 = 4, but since in the next column 2 < 8 (i.e 2 is less than 8) we lessen the 4 by 1, put down the 3 and carry 1 "On The Flag" to the 2 to make it 12.

```
7'3 5'2 -
2 7 1 8
-------
4 6 3 4
```

• To complete, say 12 − 8 = 4. Put down 4.

DIGIT SUM CHECK

```
7 3 5 2 -  ......→  8 -
2 7 1 8    ......→  0
-------            ---
4 6 3 4    ......→  8
```

CHAPTER 5 continued

SUB-TRACT-ION

USING:

"ON THE FLAG"

DIGIT SUMS

Checking results with Digit Sums is quickly done by "Casting out the Nines". In Exercise 5.3 you will be required to write digit sums to the right of the 3 number rows. Since 8 − 0 = 8 then the sum is correct.

eg 2) Here is a different example where the middle numbers are the same:

```
6 5 3 -
2 5 7
-----
4
```

• Say 6 − 2 = 4. We notice that the middle columns are the same and proceed to the next column. Since 3 < 7, we lessen the original 4 by 1, put down 3 and carry the 1 'On The Flag' to the top row 5 which becomes 15.

• Say 15 − 5 = 10, but since 3 < 7, in the next column, we reduce 10 by 1 = 9 and put ... cont.

```
6'5'3 -
2 5 7
-----
3 9 6
```

this down, also carrying 1 'On The Flag', to the top row 3 which becomes 13.
Say 13−7 = 6.
Now check with Digit Sums

6 5 3 −	......>	5 −	......>	row 1
2 5 7	......>	5	......>	row 2
3 9 6	......>	9	......>	row 3

nb 1): when a '9' appears in the 3rd row, it has the same energy as a Zero. Since 5−5=0 then the subtraction is correct.
nb 2): If row 1 is less than row 2, in its digit sums, you need to add 9 to it.

EXERCISE 5.3

1. Perform the following Subtractions working from the Left to the Right
2. Then check in two ways that your answer is correct:
a) Subtract from Right to Left, if you had learnt the 'Western' way.
b) Examine the Digit Sums, by firstly 'Casting Out the Nines'. Write the Digit Sum in pencil to the right of the 3 row of numbers. (If a 9 appears in the 2nd row, treat it as a zero).

a. 54 − 18

b. 345 − 162

c. 976 − 378

d. 4567 − 1282

e. 6245 − 1247

f. 5373 − 2838

cont.

g. 65432 −
   16754

h. 742068 −
   246363

nb: In Ex 5.3g above, there appears to be an inconsistency with the Digit Sum Check:

65432 −	⟶	2 −
16754	⟶	5
48678	⟶	6

It appears incorrect, that 2 − 5 = 6. What is needed in this case is to add 9 to the 2, then subtract 5 ie (2 + 9) − 5 = 6

PYTHAGORAS

circa 500 BC, The father of ancient Mathematike, Philosophy (Philos = Love, in Greek, + Sophia = Wisdom Goddess, ∴ meant Love of Wisdom) passed on his teachings orally; they are known only to his disciples and are recorded in 'The Golden Verses' as the surviving fragments of his works. In his Academy, there was a 5 year probation of silence, meditation, temperance, vegetarianism and secrecy regarding the teachings.

This easy to understand Sutra " BY ADDITION AND SUBTRACTION ", also known as 'Sankalana Vyavakalanābhyām', in Sanskrit, allows for rapid mental calculation. You probably are already using it, naturally.

It helps when adding or subtracting numbers like 9 or 8 or 298 which are near a Base Number eg 10 or 300.

Here are 4 examples:

eg 1. $7 + 9 = ?$
Think of the 9 as a 10, then subtract 1
$= (7+10) - 1 = 17 - 1 = 16$

eg 2. $66 + 19 = ?$
Think of the 19 as 20 then subtract 1
$= (66 + 20) - 1 = 86 - 1 = 85$

eg 3. $234 + 298$
Think of the 298 as 300 then

Chapter 5 continued

" BY ADDITION AND SUBTRACTION "

"Do not imagine that mathematics is hard and crabbed and repulsive to common sense. It is merely the etherealization of common sense".

Lord Kelvin

subtract 2
$= (234 + 300) - 2 = 534 - 2 = 532$

eg 4. $555 - 296$
Think of the 296 as 300 then add 4 $= (555 - 300) + 4$
$= 255 + 4 = 259$.

EXERCISE 5.4

Use the Sutra "By Addition and Subtraction" to solve the following sums:

a. $46 + 9 =$
b. $435 + 9 =$
c. $325 + 8 =$
d. $47 + 19 =$
e. $849 + 39 =$
f. $702 + 297 =$
g. $7279 + 498 =$
h. $120 - 9 =$
i. $230 - 8 =$
j. $796 - 19 =$
k. $361 - 28 =$
l. $491 - 47 =$
m. $852 - 297 =$
n. $7662 - 996 =$

Here is a peculiar DIGIT SUM Discovery based on Subtraction:

Take any 2 Digit number, with the ten's digit larger than the unit's digit. When you reverse those 2 digits and subtract it from the original number, look at the Digit Sum.

eg 1.

$$\begin{array}{r} 64 \\ -\ 46 \\ \hline 18 \end{array}$$

and 1 + 8 = 9

eg 2.

$$\begin{array}{r} 321 \\ -\ 123 \\ \hline 198 \end{array}$$

and 1+9+8 = 9

eg 3.

$$\begin{array}{r} 4321 \\ -\ 1234 \\ \hline 3087 \end{array}$$

and 3+0+8+7 = 9

What would happen if you didnt reverse the number but just rearranged the digits ?

DIGIT SUMS CHAPTER 5 continued

THE INDESTRUCTIBILTY OF 9

eg 4.

$$\begin{array}{r} 4231 \\ -\ 2413 \\ \hline 1818 \end{array}$$

and 1+8+1+8=9

The Explanation for this mathematical conundrum lies in the ubiquitous 9-TIMES TABLE :

9	x	1	=	9
9	x	2	=	18
9	x	3	=	27
9	x	4	=	36
9	x	5	=	45
9	x	6	=	54
9	x	7	=	63
9	x	8	=	72
9	x	9	=	81
9	x	10	=	90
9	x	11	=	99
9	x	12	=	108

Notice that as the unit's column descends "BY ONE LESS THAN THE ONE BEFORE", the ten's column rises, according to the Sutra: "BY ONE MORE".

CHAPTER 5

HOW TO MULTIPLY BY 9 BY ONLY SUBTRACTING NEIGHBOURING DIGITS!

Since Vedic Mathematics is AMBIDEXTROUS by nature, there are 2 Methods to achieve the same result:
Method 1: To ×9, subtract digits from Left ⟶ Right.
Method 2: To ×9, the converse is true ie from Left ⟵ Right.
Before we can do this, we need to know: VINCULUM:

VINCULUM

This is the concept of minus or negative numbers.
Thus $8 = 1\bar{2} = 10$ minus 2
It is read as "One Bar Two"
$88 = 1\overline{12}$ or $1\bar{1}\bar{2}$ read as
$100 - 10 - 2$
Sometimes we need to convert barred numbers to normal script eg
$3323\bar{7} = 33230 - 7 = 33223$
Notice we took the complement of 7, which is 3, and lessened the digit before it by 1. eg:
$33,\bar{2}3\bar{7} = 33030 - 200 - 7 = 32,823.$

Don't worry if you don't understand Vinculum yet, you will study it later.

Solve 3647×9 using 2 Methods of Subtraction:

METHOD 1: Left ⟶ Right

- Place a Zero on either end.

$\therefore 3647 = 036470$

- Begin from the 3, subtract in turn, the number before it.

$3 - 0 = 3$
$6 - 3 = 3$
$4 - 6 = \bar{2}$ ($\bar{2}$ = bar 2)
$7 - 4 = 3$
$0 - 7 = \bar{7}$ ($\bar{7}$ = bar 7)

- Starting from the top of the column, write the numbers in a linear fashion:

$= 33\bar{2}3\bar{7}$
$= 33030 - 200 - 7$
$= 32,823$

$\therefore 3,647 \times 9 = 32,823$
solved merely by subtractions of neighbouring digits!

continued

METHOD 2: LEFT ← RIGHT

- 3647 × 9 from Right to Left.
  = 0 3 6 4 7 0 ←
- Begin from the last Zero and subtract, in turn, the number before it

$0 - 7 = \bar{7}$
$7 - 4 = 3$
$4 - 6 = \bar{2}$
$6 - 3 = 3$
$3 - 0 = 3$ ↑

- To write this linearly, start from the bottom of the column

$= 33,\bar{2}3\bar{7}$
$= 32\,823$

EXERCISE 5.5

Multiply the following nos. by 9, using both methods:

a. 73
b. 123
c. 3,846
d. 6,666

cont... Chap. 5

X 9 BY DIGITAL SUBTRACTION

Ambi-Dexterity using 2 methods.

BASE	Q	R	BAR Number		
10 ÷ 9	1	1	$9 = 1\bar{1}$	or	10 − 1
20 ÷ 9	2	2	$18 = 2\bar{2}$	or	20 − 2
30 ÷ 9	3	3	$27 = 3\bar{3}$	or	30 − 3
40 ÷ 9	4	4	$36 = 4\bar{4}$	or	40 − 4
50 ÷ 9	5	5	$45 = 5\bar{5}$	or	50 − 5

Perhaps one explanation why multiplication by 9 can be achieved via digital subtractions lie hidden in Nine's relationship to its Bases like 10, 20, 30 etc. The Discovery, in the above chart is that, upon division by 9, the Remainder (R) is EQUAL to its Quotient (Q).

Another curiosity about ×9 and Subtraction of Digits was explored in Exercise 2.11 eg 123 × 9 = 1230 − 123 = 1107 !

APPENDIX

SPECIAL DIVISION BY 9

EXAMPLES

Here is a excerpt from Book 3 on Division, especially fun and easy for 10 - 12 year olds. It is included in this book on Digit Sums as it involves an unusual though simple addition of neighbouring digits!

eg 1: Divide 1023 by 9.

The setting out looks like this:

```
9)102 ¦ 3
  113 ¦r6
```

The dotted line is placed before the last digit of the dividend as a marker for the Remainder (r).

```
9)102 ¦ 3
  ↓↗   ¦
  1    ¦
```

The initial 1 is dropped into the answer and is added to the Zero diagonally across and another 1 is put down.

```
9)102 ¦ 3
  11↗  ¦
```

This second 1 is added to the 2 diagonally across and 3 is put down.

```
9)102 ¦ 3
  113↗ ¦
```

Then this 3 is added to the top 3 to give a remainder of 6.

```
9)102 ¦ 3
= 113 ¦r6
```

QUESTION: What happens when 2 added digits excede 9?

eg 2: Divide 713 by 9.

```
9) 71 ¦ 3
   ↓  ¦
   78 ¦r11
```

We need to subtract 9 from the remainder of 11 and also add 1 (lot of 9) to the preceding 8.

```
9) 71 ¦ 3
 = 79 ¦r2
```

eg 3: Divide 6145 by 9

```
9) 614 ¦ 5
   ↓     ↗
   6 7 11
 = 6 8 2 ¦r7
```

By taking 9 from the 11, 1 must be added to the 7 to make 8.

EXERCISE A.1

Divide the following numbers by 9, by the method just shown that adds diagonally neighbouring digits. Let 'r' = the remainder.

a. 62
b. 123
c. 241
d. 12,302
e. 343
f. 6,012
g. 40,211
h. 20,000
i. 31,085
j. 60,000
k. 70,000
l. 2,468
m. 123,456
n. 1,234,567

SPECIAL ÷ BY 9

BUDDHA KNOWS MATHS.

Before Buddha's spiritual enlightenment in 600 BC and prior to his marriage to Gopa, he had to pass a test to win her hand, as she had 500 suitors.

The examinations included Music, Archery and Mathematics. Having vanquished all his rivals, except Arjuna, a showdown or challenge ensued to determine whose knowledge of maths and science was the greater.

Buddha chose a very complicated operation in Arithmetic. He proceeded to describe the number of 'Primary Atoms' which placed end to end would form a line with length equal to the ancient Hindu mile. He recited: "Seven Primary Atoms make a very minute grain of dust. Seven of these grains of dust whirled up by the wind..." He continued. When he reached the length of a mile, all the 7's listed by Buddha yielded a product of about 50 digits!

(Recorded in a sacred Buddhist text: "LALITAVISTARA").

MAGIC SQUARES AND THE SUTRAS

As a concluding exercise, I would like to offer some insight into the internal harmony of Magic Squares and their relationship with some of the Sutras, hidden in their Digit Sums and other peculiarities. This passionate study of Magic Squares, for 24 years, enabled me to fully comprehend and appreciate the inherent visual patterns in Vedic Mathematics. The page on your right has been reproduced, in full, from "JOIN – THE – DOTS", 1999, as an example of "Mathematical Art".

EXERCISE A.2

Connect all the Pairs that have a sum of 26, as explained in the box on your right titled "ACTIVITY":

THE BALANCE OF THE PAIRS

11	24	7	20	3
4	12	25	8	16
17	5	13	21	9
10	18	1	14	22
23	6	19	2	15

Fig 5a

PROPERTIES

- All Pairs balanced around the central cell add up to 26
- The central Cell is '13' which is half of 26.
- This number '26' is the sum of the first number (1) + the last number (25).
- This Balancing of the Pairs indicates that there is an internal order existing amidst the apparent disorder.

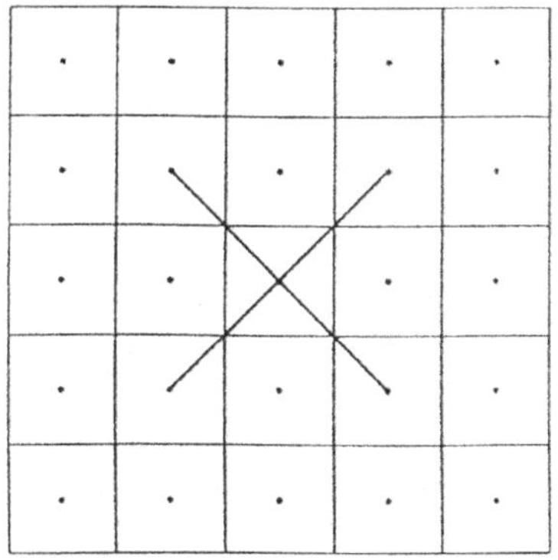

Fig 5f

ACTIVITY

In the Space of Fig 5f connect all the Pairs of numbers, of Fig 5a above, that have a sum of 26. Each dot represents a number.
Two pairs have already been drawn for you. They are: 8 and 18 = 26
12 and 14 = 26

As usual, refer to the end chapter 'SOLUTIONS' to check your results.

MAGIC SQUARE of 5×5 REDUCED TO ITS DIGIT SUMS + OTHER SUTRAS

2	6	7	2	3
4	3	7	8	7
8	5	4	3	9
1	9	1	5	4
5	6	1	2	6

When the double digits of the Magic Sq. of 5×5 are reduced to single digits, you will realize that the exercise just performed of linking all the Pairs adding up to 26 was, in effect, another Sutra at play. You really just linked all the Digit Sums that add up to 8. This Combination of Pairs is called, in the Sutraic model: "DWANDWA YOGA" or "DUPLEX". This Sutra is all about Squared numbers (5×5) and embraces any process of combining numbers, like the Digit Sums adding to 8; a union of 2 energies. You will learn a lot more about 'Duplex' in the next book on "X" to solve problems like "345^2".

1	2	3	4	5
6	7	8	9	10
11	12	13	14	15
16	17	18	19	20
21	22	23	24	25

In Chapter 3, Ex 3.6, you studied Natural Squares. When you compare the Magic Sq. of 5×5 with its Natural Square and study the 'Similarities and Differences' it brings into play another Sutra called "CALANA KALANA", also: Unity and Diversity. eg: If you add either of the 2 diagonals of the Natural Square of 5×5 above, it is equal to, and determines, the Magic Sum of 65 of the Magic Sq. of 5×5!

Also, if you interchange the top row of the Magic Sq. of 5×5 with any other row, or interchange the left-most column with the right-most column, you do not lose the integrity of the Magic Sq. This TransPosition (French: 'trans' = across + 'pose' = place) of rows and columns relates to the Sutra: 'PARAVARTYA YOJAYET' = TRANSPOSE + APPLY!

Answers

SOLUTIONS

Exercise 1.1

a. $1 = 1 \times 1 = 1^2$

b. $1+3 = 2 \times 2 = 2^2$

c. $1+3+5 = 3 \times 3 = 3^2$

d. $1+3+5+7 = 4 \times 4 = 4^2$

e. $1+3+5+7+9 = 5 \times 5 = 5^2$

f. $1+3+5+7+9+11 = 6 \times 6 = 6^2$

g. $1+3+5+7+9+11+13 = 7 \times 7 = 7^2$

h. $1+ \longrightarrow +15 = 8 \times 8 = 8^2$

i. $1+ \longrightarrow +17 = 9 \times 9 = 9^2$

j. $1+ \longrightarrow +19 = 10 \times 10 = 10^2$

k. $1+ \longrightarrow +21 = 11 \times 11 = 11^2$

l. $1+ \longrightarrow +23 = 12 \times 12 = 12^2$

CONCLUSION: The addition of the next sequential odd number generates the next Squared Number.

Exercise 1.2

a. $1 = 1 \times 1 \times 1 = 1^3$

b. $3+5 = 2 \times 2 \times 2 = 2^3$

c. $7+9+11 = 3 \times 3 \times 3 = 3^3$

d. $13+15+17+19 = 4 \times 4 \times 4 = 4^3$

e. $21+23+25+27+29 = 5 \times 5 \times 5 = 5^3$

f. $31+33+35+37+39+41 = 6 \times 6 \times 6 = 6^3$

g. $43+45+ \longrightarrow +55 = 7 \times 7 \times 7 = 7^3$

Exercise 1.3

a. 24

b. 112

c. 2,106

d. 32,104

e. 201,323

Exercise 2.1

a. 7

b. 4

c. 3

d. 5

e. 5

f. 4

g. 8

Exercise 2.2

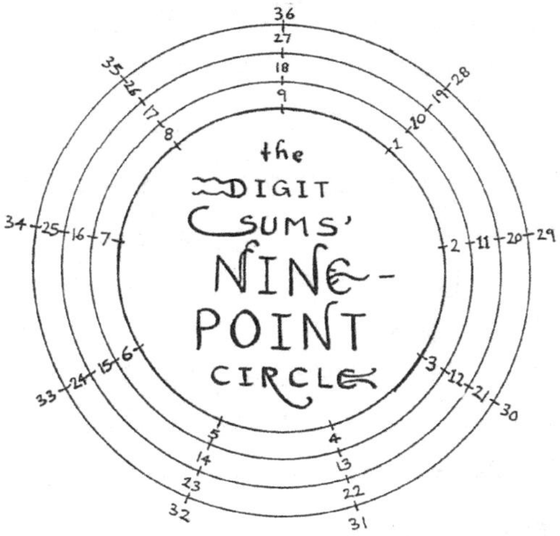

SOLUTIONS

- Digit Sums of Ray 1 (1, 10, 19, 28) = 1, 1, 1, 1
- The next 5 numbers on Ray 1 = 37, 46, 55, 64 and 73
- The Digit Sums for any of the 9 Rays are the same as the number it starts with. eg: Ray 3 has the numbers: 3, 12, 21, 30 etc and their Digit Sums are all 3 !

Exercise 2.3

a. 4
b. 3
c. 3
d. 6
e. 4
f. 3
g. 7
h. 1

Exercise 2.4

a. 5
b. 2
c. 3
d. 2
e. 7
f. 3

Exercise 2.5

a. $3+4+5 = 3$ $\quad 6+7+8 = 3$
$2+3+3+9+1+0 = 9$
Since $3 \times 3 = 9$ = Digit Sum of the Product (9), then it is correct.

b. $6+7+5+1 = 1$ $\quad 3+6+9+4 = 4$
$2+4+9+3+8+1+9+4 = 4$
Since 1×4 = Digit Sum of the Product (4), then it is correct

c. $4+5+2 = 2$ $\quad 6+3+2+7+4 = 4$
$2+8+5+9+9+8+3+8 = 7$
Since $2 \times 4 \neq 7$ then it is incorrect.
($\neq$ means "does not equal").

Exercise 2.6

a. 231
b. 385
c. 462
d. 572
e. 693
f. 781
g. 891
h. 990

Exercise 2.7

a. 418
b. 308
c. 539
d. 616
e. 726
f. 869
g. 957
h. 1,089

Exercise 2.8

a. 1,353
b. 26,983
c. 45,782
d. 583,264
e. 269,643
f. 1,235,938

Exercise 2.9

a. Yes c. No
b. Yes d. No

Exercise 2.11

a. 12,221 c. 1,356,531
b. 13,431 d. 135,787,531

Exercise 3.1

The D.S.S.P (or Digital Sum Series Pattern) for 1 → 10 Cubed is
DSSP = 1, 8, 9 Periodicity (P) = 3

Exercise 3.2

D.S.S.P for TriΔngular Numbers
= 1, 3, 6, 1, 6, 3, 1, 9, 9 (P) = 9

Exercise 3.3

D.S.S.P for Squared Numbers
= 1, 4, 9, 7, 7, 9, 4, 1, 9 (P) = 9

Exercise 3.4

D.S.S.P for Last Digit of the Squared Numbers
= 1, 4, 9, 6, 5, 6, 9, 4, 1, 0 (P) = 10

Exercise 3.5

D.S.S.P for Magic Square Constants
= 6, 7, 2, 3, 4, 8, 9, 1, 5 (P) = 9

Exercise 3.6

D.S.S.P for Natural Squares
= 1, 1, 9 (P) = 3

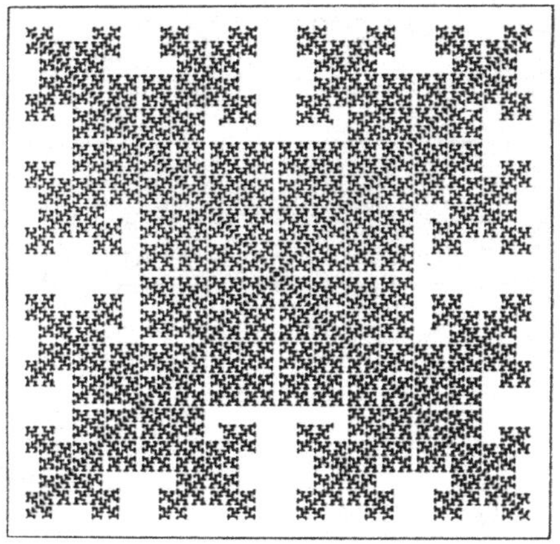

cont… ANSWERS : Exercise 3.7

The **DIGITAL SUMS** of the famous **FIBONACCI SEQUENCE** has a **period** (or predictable circulation) of **24**
i.e: the chain of single digits recurs every 24th digit.

Here are the **Digital Sums** of the first 24 Fibonacci Numbers:
1, 1, 2, 3, 5, 8, 4, 3, 7, 1, 8, 9, 8, 8, 7, 6, 4, 1, 5, 6, 2, 8, 1, 9

The obvious symmetry is based on a **COMPLEMENTARY CYCLE OF NINES** shown below.

1	**1**	**2**	**3**	**5**	**8**	**4**	**3**	**7**	**1**	**8**	**9**
8	**8**	**7**	**6**	**4**	**1**	**5**	**6**	**2**	**8**	**1**	**9**

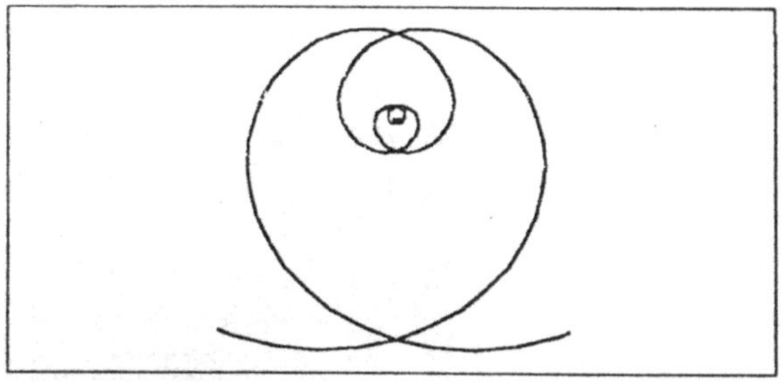

The Golden Mean spirals traveling in opposite directions create the image of the heart.

For more info, see:
1) "The Magic of NINE in Vedic Mathematics". Jain, pub. 2001.
2) "The Book of PHI". Jain, to be published in 2002.

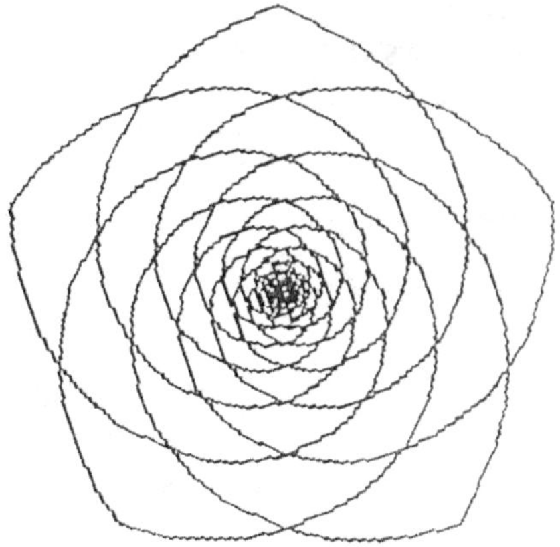

Exercise 3.8

D.S.S.P for "Pyramid of Threes" = 9 and (P) = 1

33^2 = ... 1,089
333^2 = ... 110,889
$3,333^2$ = ... 11,108,889
$33,333^2$ = ... 1,111,088,889
$333,333^2$ = ... 111,110,888,889
$3,333,333^2$ = . . . 11,111,108,888,889
$33,333,333^2$ = 1,111,111,088,888,889
$333,333,333^2$ = 111,111,110,888,888,889

Exercise 3.9

Questions (Q) on Pascal's / Halayudha's TriAngle :

Q 1) Sum of Rows
= 1, 2, 4, 8, 16, 32, 64, 128

Q 2) Each following term is double the preceding term ie they are based on the Powers of 2.
eg: $2^0, 2^1, 2^2, 2^3, 2^4, 2^5, 2^6, 2^7$

Q 3) $2^8 = 256$, $2^9 = 512$, $2^{10} = 1,024$
$2^{11} = 2,048$, $2^{12} = 4,096$

(SOLUTIONS)

Q 4) $\frac{1}{1}$ $\frac{2}{2}$ $\frac{4}{4}$ $\frac{8}{8}$ $\frac{7}{7}$ $\frac{5}{5}$ 1

Q 5) The Digit Sum of the first 8 rows of Pascal's/Halayudha's Δ was not enough information to determine recursion. After the first 6 digit sums of 1, 2, 4, 8, 7, 5 the next 2 did repeat which indicated possible recursion. After extending to 13 rows, the recursion of 6 circulating digits was verified.

Q 6) D. S. S.P for Pascal's / Halayudha's TriAngle
= 1, 2, 4, 8, 7, 5 (P) = 6

Exercise 3.10

11^6 = 1,771,561
11^7 = 19,487,171
11^8 = 214,358,881

Exercise 3.11

1	2	3	4	5	6	7	8	9
2	4	6	8	1	3	5	7	9
3	6	9	3	6	9	3	6	9
4	8	3	7	2	6	1	5	9
5	1	6	2	7	3	8	4	9
6	3	9	6	3	9	6	3	9
7	5	3	1	8	6	4	2	9
8	7	6	5	4	3	2	1	9
9	9	9	9	9	9	9	9	9

The Vedic Square

Exercises 3.12 and 3.13

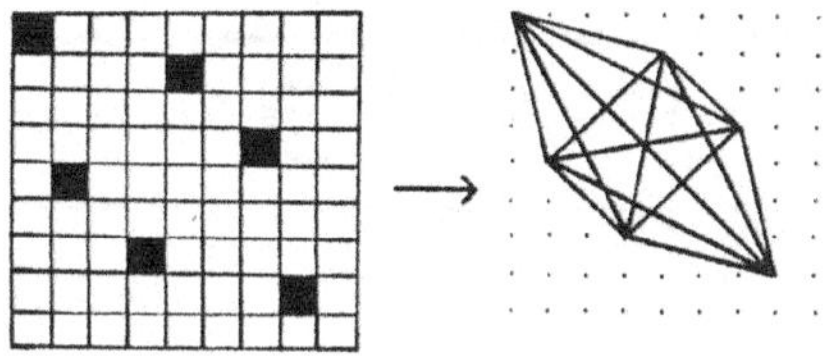

"ALL THE ONES"

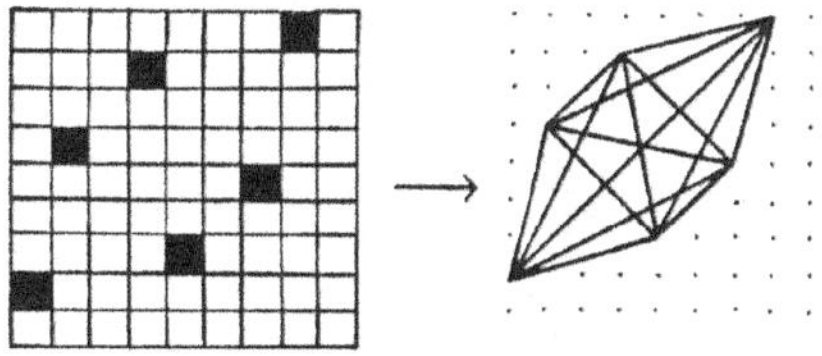

"ALL THE EIGHTS"

You have just created the Atomic Structure of Rutile (Titanium Dioxide) a crystal found in white beach sands where black streaks appear. It has been the centre of an ecological controversy in Australia when it was sand-mined in the 1980s. This crystal can be seen as two pyramids joined base-to-base forming, what healers would call, a double-terminating octahedral crystal.

Exercise 3.14

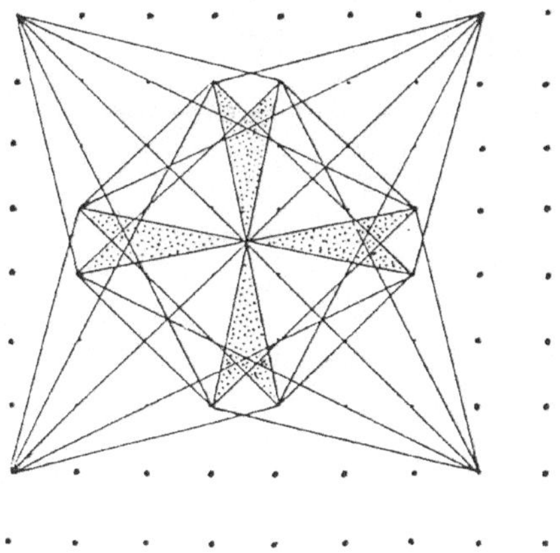

Combination of All The Ones and All The Eights to form: PLATINUM CRYSTAL!

Exercise 3.15

a. Yes
b. Yes
c. No
d. No
e. Yes
f. No
g. Yes
h. No

Exercise 3.16

a. Yes
b. Yes
c. No
d. Yes
e. Yes
f. Yes

Exercises 3.17 and 3.18

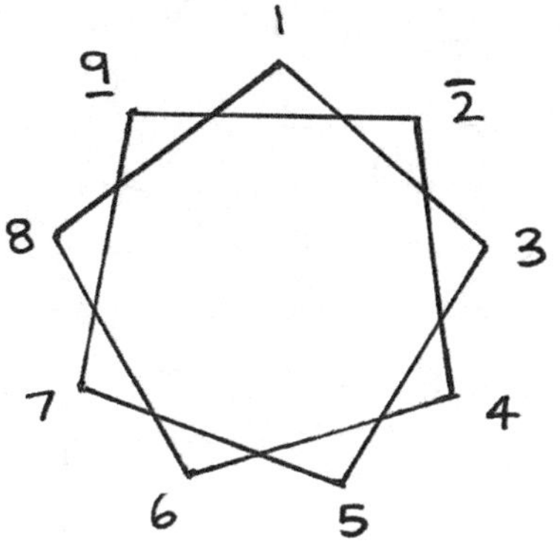

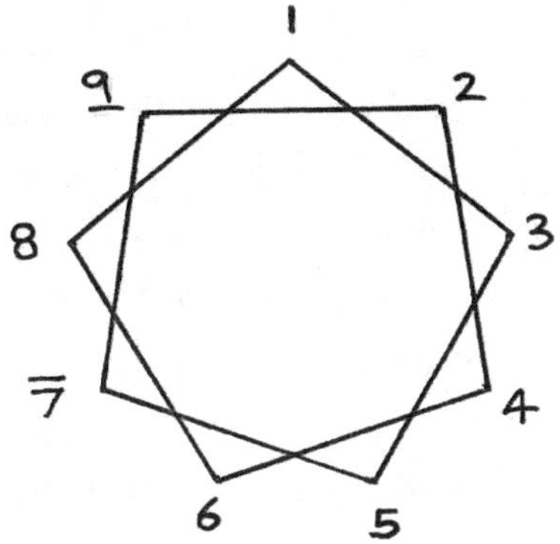

Q) The Pattern for Row 7 of the Vedic Square is drawn in the REVERSE direction of the Pattern for Row 2. Also, 2 and 7 are Complements or Number Pairs of 9.

The Book of DIGIT SUMS

Exercises 3.19 and 3.20

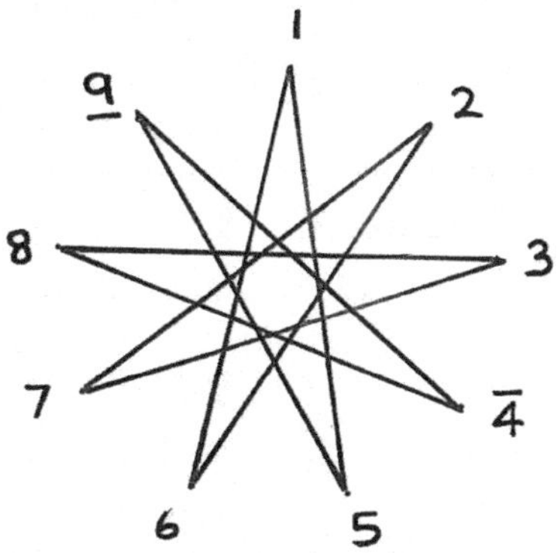

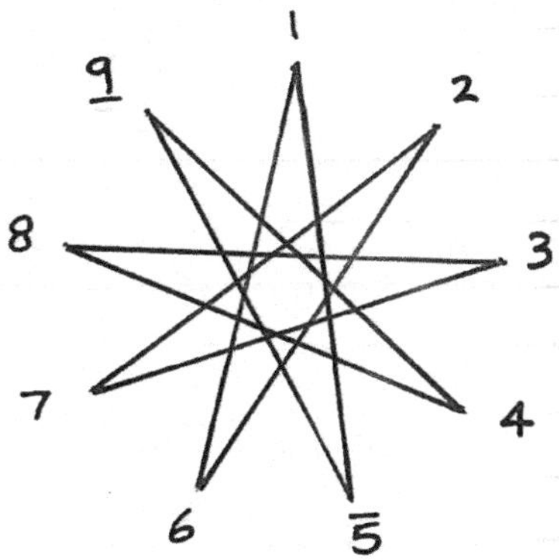

Q1) This shape is called The ENNEAGRAM (from the Greek word "Ennéa" = Nine).

Q2) The Pattern for Row 5 is drawn in the REVERSE direction of the Pattern for Row 4. Also, 4 and 5 are Complements of 9.

Exercise 4.1

a. 19 = 2
b. 69 = 7
c. 139 = 14
d. 9 = 09 = 1
e. 7 = 7×7 = 49 = 5
f. 13 = 13×3 = 39 = 4
g. 17 = 17×7 = 119 = 12
h. 11 = 11×9 = 99 = 10
i. 27 = 27×7 = 189 = 19
j. 3 = 3×3 = 09 = 1

Exercise 4.2

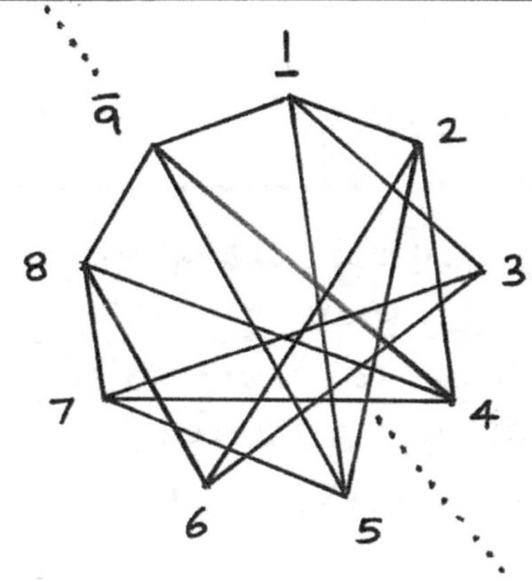

There exists in this diagram of 18 cyclic digits a symmetry about the axis (shown as the dotted line), which means it can fold over or reflect onto itself. This reflection line or axis is symbolised by the "midway slash" in the decimal $\frac{1}{19}$ = $\cdot\dot{0}52631578 / 947368421$ also determined by Denominator minus Numerator or 19-1 = 18 and is the half-way point or magical gateway where the Complements of 9 are reeled off mentally to complete the answer.

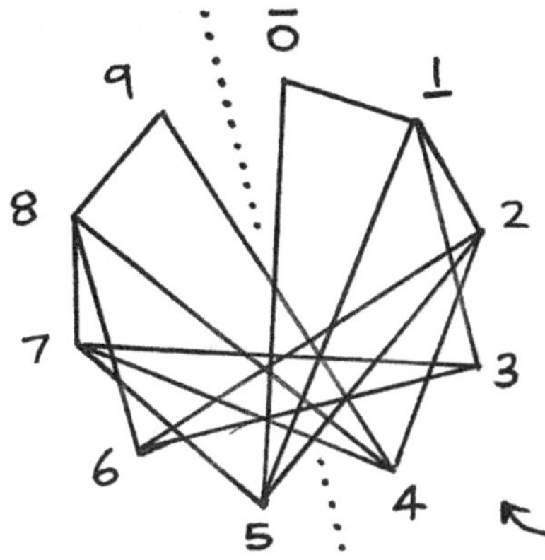

To perfect this diagram, construct a 10-Point Circle to distinguish the Zero from the Nine and draw the 18 digit cycle again, placing a bar over the Zero ($\bar{0}$) to indicate where the decimal begins and a bar below the 1 ($\underline{1}$) to indicate where it ends. (Notice also that sometimes, when connecting 2 numbers, the line has already been drawn, like from 3 → 6, so you merely go over it or ignore it and move to the next line from 6 → 8).

Similar to the 9-Point Circle, the 10-Point Circle expression of $\frac{1}{19}$ also has an axis of symmetry.

Exercise 5.1

a. 68
b. 140
c. 768
d. 624
e. 80235

Exercise 5.2

a. 60
b. 112
c. 69
d. 140
e. 220
f. 22,220
g. 25,716

Exercise 5.3

a. 36	d. 3,285	g. 48,678
b. 183	e. 4,998	h. 495,705
c. 598	f. 2,535	

Exercise 5.4

a. 55	f. 999	k. 333
b. 444	g. 7,777	l. 444
c. 3,333	h. 111	m. 555
d. 66	i. 222	n. 6,666
e. 888	j. 777	

Exercise 5.5

a. Method 1

$73 \times 9 = 0730$

$= 7-0 = 7$ ↓
$= 3-7 = \bar{4}$
$= 0-3 = \bar{3}$

$= 7\overline{43}$
$= 700 - 43$
$= 657$

Method 2

$0-3 = \bar{3}$ ↑
$3-7 = \bar{4}$
$7-0 = 7$

$= 7\overline{43}$
$= 700 - 43$
$= 657$

b. Method 1

$123 \times 9 = 01230$

$1-0 = 1$ ↓
$2-1 = 1$
$3-2 = 1$
$0-3 = \bar{3}$

$= 111\bar{3}$
$= 1110 - 3$
$= 1107$

Method 2

$0-3 = \bar{3}$ ↑
$3-2 = 1$
$2-1 = 1$
$1-0 = 1$

$= 111\bar{3}$
$= 1110 - 3$
$= 1107$

c. Method 1

$3846 \times 9 = 038460$

$3-0 = 3$ ↓
$8-3 = 5$
$4-8 = \bar{4}$
$6-4 = 2$
$0-6 = \bar{6}$

$= 35\bar{4}2\bar{6}$
$= 35{,}020 - 400 - 6$
$= 34{,}614$

Method 2

$0-6 = \bar{6}$ ↑
$6-4 = 2$
$4-8 = \bar{4}$
$8-3 = 5$
$3-0 = 3$

$= 35\bar{4}2\bar{6}$
$= 35{,}020 - 400 - 6$
$= 34{,}614$

d. pto

d. $6666 \times 9 = 066660$

Method 1	Method 2
$6-0 = 6$ ↓	$0-6 = \bar{6}$ ↑
$6-6 = 0$	$6-6 = 0$
$6-6 = 0$	$6-6 = 0$
$6-6 = 0$	$6-6 = 0$
$0-6 = \bar{6}$	$6-0 = 6$
$= 6000\bar{6}$	$= 6000\bar{6}$
$= 60,000 - 6$	$= 60,000 - \bar{6}$
$= 59,994$	$= 59,994$

Exercise A.1

a. 6 r 8
b. 13 r 6
c. 26 r 7
d. 1,366 r 8
e. 38 r 1
f. 668 r 0
g. 4,467 r 8
h. 2,222 r 2
i. 3,413 r 8
j. 6,666 r 6
k. 7,777 r 7
l. 274 r 2
m. 13,717 r 3
n. 137,174 r 1

Exercise A.2

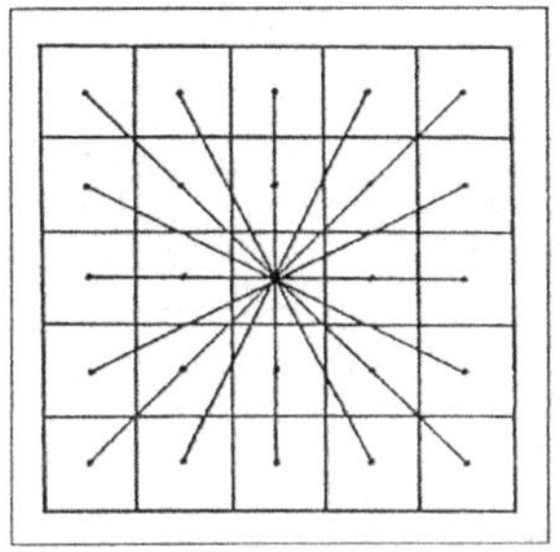

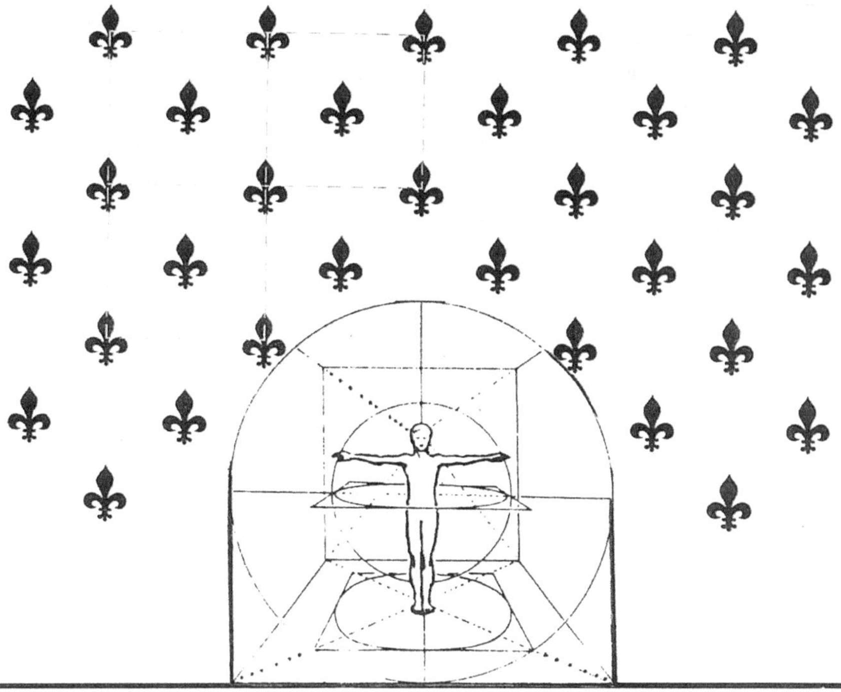

BIBLIOGRAPHY

1. VEDIC MATHEMATICS by Bharati Krsna Tirthaji Pub. by Motilal Banarasidas 1965 www.mlbdbooksuk.com ISBN 0-8426-0967-9
2. FIGURING. THE JOY OF NUMBERS by Shakuntala Devi. 1986. Orient Paperbacks. ISBN 81-222-0038-9
3. MAHARISHI MAHESH YOGI (Transcendental Meditation Founder). www.maharishi.co.uk
4. THE COSMIC COMPUTER by Kenneth Williams and Mark Gaskell 1997 www.vedicmaths.org.uk
5. Barbara Salmon www.users.waitrose.com/~dabsalmon
6. THE LANGUAGE OF PATTERN by Keith Albarn + Jenny Smith, 1974
7. MATHEMATICS AS KNOWN TO THE VEDIC SAMHITĀS by M.D. Pandit, 1993
8. BHĀGĀVĀD GITĀ. AS IT IS translated by Bhaktivedanta Swami Prabupada 1984

9. HOW MATHS WORKS by Carol Vorderman. A DK book.
10. THE MATHEMATICS OF THE COSMIC MIND 1970 by L. Gordon Plummer
11. ALPHABET OF THE HEART by Dan Winter's research inspired and adapted from Stan Tenen who created the original 3-D Models. 1994 www.danwinter.com www.meru.org (S. Tenen). Please review both websites to understand the current controversy. The above book is banned and out-of-print!
12. OCCULT CHEMISTRY C.W. Leadbeater and Annie Besant. (Investigations by clairvoyant magnification into the structure of the atoms of the periodic table). 3rd Ed. 1951.
13. MATHS OR MAGIC by Joseph Howse 1976.

ACKNOWLEDGEMENTS of GRAPHICS USED

PAGE 1 : The word "Earth" joined like a necklace 7 times now reads as : "Heart" 7x . The Earth Heart connection. The Star in the middle is the Magic Square of 3x3 superimposed 4 times at angles of 45° : my LOGO .

PAGE 3 : Extended 'Fruit-of-Life' Pattern by Drunvalo Melchizedek from 'The Ancient Secret of the Flower of the Life . Volume 2 . 2000.

PAGE 4 : Magic Squares on The Tree of Life depicting Qabalistic planes of consciousness or planetary levels or subtler dimensions of existence.

PAGE 8 : 'Unknown Man : The Mysterious Birth of a New Species' from Yatri, p 208 .

PAGE 9 : 'Photo of a Water Drop' by Martin Dohrn, taken from 'How Maths Works' by Carol Vorderman.

PAGE 10 : 'Self-Confidence' photo 'Here and Now' Osho mag, 2001.

PAGE 11 : Photo of author of 'Vedic Mathematics', latest edition.

PAGE 13 : Painting of Shiva 'Lord of the Dance' from an Indian postcard.

PAGE 14 : text on pp 11-14 from www.vedicmaths.org.uk/group_files/newsletter, an excellent source of information by Kenneth Williams and friends.

PAGE 15 : ibid + page 16 .

PAGE 17 : The Golden Flower, ancient Chinese meditative art .

PAGE 21 : Odds + Evens from a chapter "Seven, The Number of Creation"; title of book unknown .

PAGE 28 : Shri Yantra or Sri Chakra, 9 interlacing triangles (relating to the 9 wifes of Shiva or feminine aspects), and having 43 triangular compartments zoned for goddess worship .

PAGE 39 : Graphics and text by Dan Winter and Stan Tenen !?? See 'Alphabet of the Heart' in Bibliography for more details on the controversial copyright !

PAGE 47 : ibid

PAGE 51 : text from Barbara Salmon's website, see Bibliography

Acknowledgements

PAGE 58 : diagrams on left from 'Language of Pattern' by Keith Albarn and Jenny Smith
PAGE 59 : 'The 4 Angels of 9' by Jain. I had independently received and worked with the Vedic Square since age 21.
PAGE 62 : segment of 'Olivetti' calligraphy "systems pour l'information".
PAGE 63 : Islamic brickwork, 9 point-star tessellations or tilings.
PAGE 64 : Buddha in Divine Proportion from 'The Power of Limits' by G. Doczi.
PAGE 66 : Fairy art, artist unknown.
PAGE 67 : Renaissance era border
PAGE 71 : Buddha in the Phi Ratio or Golden Rectangle from 'The Power of Limits : Proportional Harmonies in Nature, Art and Architecture' by György Doczi 1981, Shambhala Publications ISBN 0-87773-194-2
PAGE 74 : Girl playing harp, Artist unknown.
PAGE 75 : Border in temple form, Academicism, Munich 1839.
PAGE 76 : label from Prothiaden Medicine: Dolhiepin Hydrochloride B.P
PAGE 83 : Illustration of Pythagoras from "The History of Philosophy' by Thomas Stanley, 1655-1662.
PAGE 89 : Buddha story adapted and copied in part from Shakuntala Devi's "The Book of Numbers".
PAGE 95 : Fractal Art. Best definition is like a "Branch" that is self-similar to the "Tree". Each "Twig" or smaller segment of the square is self-similar to the larger square.
PAGE 96 : both images from Dan Winter's book "Alphabet of the Heart"
PAGE 99 : Scroll design from a rare 1828 Christian Bible in the Author's Library.
PAGE 101 : Earth Repair Foundation Stickers by Frank Scarf of N.S.W
PAGE 107 : 3-Dimensional Flower-of-Life, symbol of Melchizedek Order or Lords of Light (plagarized by Alton !).
PAGE 110 : 'The Anatomy of Internal Viewing' from 'Hands of Light' by Barbara Ann Brennan, p 164.

ORDER FORM

F.R.E.E.D.O.M.S.

FOR RESEARCH EXPRESSING ESSENTIAL DATA OF MAGIC SQUARES

E-mail: jainmagic42@hotmail.com

TITLE	PUB	PP	GMS	WP	RRP	QTY	TOTAL
The BOOK of MAGIC SQUARES Volume 1	1990	164	600	$25	$35		
The BOOK of MAGIC SQUARES Volume 2	1996	160	600	$25	$35		
The BOOK of MAGIC SQUARES Volume 3	2000	174	600	$25	$35		
JOIN-THE-DOTS Book 2	1999	56	230	$15	$25		
The MAGIC of 9 in VEDIC MATHEMATICS	2001	124	380	$25	$35		
The BOOK of PHI	2001						
In The NEXT DIMENSION	2002						
DECAL (LARGE) M.SQ. of 4×4	1992		85	$ 8	$ 16		
DECAL (SMALL) M.SQ. of 4×4 and 8×8	1992		25	$ 4	$ 8		
The VEDIC MATHS CURRICULUM ...GLOBAL SCHOOL	2001			$25	$35		
						TOTAL $	....

POSTAGE AND HANDLING CHARGES

Up to $40 order $3.00
$40 to $100 $8.00
Over $100 Free
} in Australia.
International mail - air freight charges apply)

JAIN
MAGIC SQUARE STUDIO
777 LEFT BANK Rd MULLUMBIMBY CREEK, 2482, Australia
Phone/Fax: 61 2 6684 4409
Website : www.linknet.com.au/jain

Something To Think About

If we could shrink the earth's population to a village of precisely 100 people, with all the existing human ratios remaining the same, it would look something like the following.

There would be:
57 Asians
21 Europeans
14 from the Western Hemisphere, both north and south
8 Africans
52 would be female
48 would be male
70 would be non-white
30 would be white
70 would be non-Christian
30 would be Christian
89 would be heterosexual
11 would be homosexual

6 people would possess 59% of the entire world's wealth and all 6 would be from the United States.

80 would live in substandard housing
70 would be unable to read
50 would suffer from malnutrition
1 would be near death; 1 would be near birth
1 (yes, only 1) would have a college education
1 would own a computer

When one considers our world from such a compressed perspective, the need for acceptance, understanding and education becomes glaringly apparent. The following is also something to ponder...

If you woke up this morning with more health than illness...you are more blessed than the million who will not survive this week.

If you have never experienced the danger of battle, the loneliness of imprisonment, the agony of torture, or the pangs of starvation... you are ahead of 500 million people in the world.

If you can attend a church meeting without fear of harassment, arrest, torture, or death...you are more blessed than three billion people in the world.

If you have food in the refrigerator, clothes on your back, a roof overhead and a place to sleep...you are richer than 75% of this world.

If you have money in the bank, in your wallet, and spare change in a dish some place ... you are among the top 8% of the world's wealthy.

If your parents are still alive and still married ... you are very rare, even in the United States and Canada.

If you can read this message, you just received a double blessing in that someone was thinking of you, and furthermore, you are more blessed than over two billion people in the world that cannot read at all. Someone once said: "What goes around comes around".

If you can do just one really nice thing for someone else today and everyday, your life will become more special than you could possibly imagine. Enjoy it!

☺

YOU DON'T HAVE TO BE A MATHEMATICIAN TO DO THIS COURSE!

You will be introduced to the first principles of Vedic Mathematics; this will also allow a parent to teach their child.

All material taught will be taken from JAIN'S book "VEDIC MATHS FOR THE AUSTRALIAN CURRICULUM" Subtitled "HOW DO I TEACH MY CHILD VEDIC MATHEMATICS?"

This covers the basic curriculum for primary and secondary schools.

ABOUT THE TEACHER

JAIN is the author of six self-published books on Sacred Geometry which he has actively been researching and teaching over 22 years. Topics include:

- Magic Squares and their Atomic Artforms.
- The Vedic Square and it's Digital Sums.
- The Divine Proportion or The Mathematics of Nature, also known as The PHI Ratio or The Golden Measure.
- The five Platonic Solids
- Pythagorean Mystery School
- Vedic Mathematics.

He is currently writing a series of booklets for the Australian Curriculum on Vedic Mathematics.

As an Authorised School Performer of a travelling show called 'Mathemagics', JAIN has taught thousands of children in Australia, the wonder of the pure mathematical principles relating to Atomic Art and Discovery.

He works with Gifted and Talented children and lectures at Maths Camps, Universities and Maths Conferences.

This course will be held in the venue of JAIN'S beautiful rainforest studio over-looking the pristine Mullumbimby Creek, 30 kms and half an hour inland of the Byron Bay coast.

7 DAY WORKSHOP

VEDIC MATHEMATICS

"TEACHER TRAINING"

UNDER THE AUSPICES OF

JAIN

SUNDAY 30TH SEPTEMBER
TO
SATURDAY 6TH OCTOBER,
2001

PH : 02 6684 4409

ABOUT THE COURSE

This course is based on the book
"VEDIC MATHEMATICS"
by Swami Bharati Krishna Tirthaji,
late Shankacharya (like a Dalai Lama)
of Puri Govardhan Math (1884 - 1960).

There are 16 simple Sanskrit Sutras (or formulae) which solve all known mathematical problems in the branches of Arithmetic, Algebra, Geometry and Calculus. These 16 Sutras and a few corollaries are easy to understand, easy to apply and easy to remember.

This **Vedic One-Line Mental Arithmetic** is very helpful in stimulating Modern Mathematicians to adopt its simplicity and quickness. Once initiated to these Vedic rules, students of all ages will appreciate and enjoy the subject.

The aim of this Course is to bring this long-hidden treasure trove of intelligent mathematical knowledge within easy reach of everyone who wishes to obtain it and benefit by it

Day begins punctually at 9am with a short meditation,
followed by 2 morning lectures, we then break for a delicious lunch
and then 2 afternoon lectures.
The day completes at 4pm.

VILOKANAM

"BY MERE INSPECTION"

This 7 Day Course is not to make obsolete the present Mathematical Curriculum, but rather to enhance it.
By relying on modern electronic calculators for the next 20 years or so, without cultivating mental calculations which build the "Mental Muscle", will sadly produce a generation of young adults whose brain will have undoubtedly atrophied!
Vedic Mathematics promotes clarity and sharpness of mind as well as improving memory skills and self-confidence, by awakening the Pineal Gland (or Third Eye) to clairvoyantly "see" or intuitively know the predicted answer.
There is a key Sutra called **Vilokanam** which means "By Mere Inspection".

A child can know the Cube Root of 35,937 to be equal to 33, By Mere Observation! No working out, no pen or paper.
Mathematics without Tears.

It is an infallible Right Brain (Feminine) application of Learning based on Pattern Recognition.

REGISTRATION

Name: ____________________

Address: ____________________

Email: ____________________

www: ____________________

Phone/fax: ____________________

Qualifications: ____________________

____

Please enrol me as a participant in the 7 day Vedic Mathematics "Teacher Training" workshop

I enclose
☐ cheque
☐ money order
to the value of $1,000

I know this includes catered lunches and morning and afternoon teas
(no credit card facilities available)

Please fill in the above form, enclose payment and post to:

JAIN
777 LEFT BANK ROAD
MULLUMBIMBY CREEK.
NSW 2482

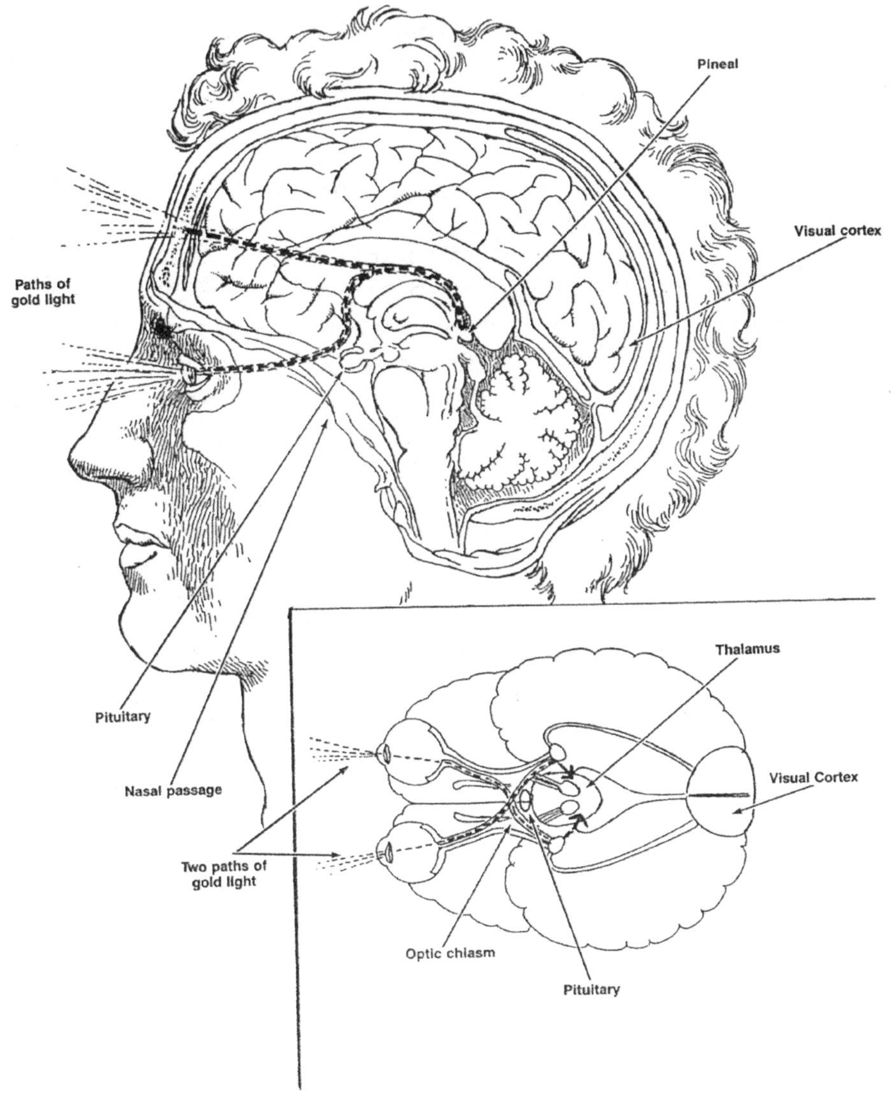

The Anatomy of Internal Viewing

NEXT ISSUE

Notice the "Cross-Over" of the optic nerves connecting with the complementary halves of the brain. This "X" is actually the shape of one of the Sutras called "Urdhva Tiryagbhyām" or "Vertically and Crosswise". In the neXt issue, Book 2, which will be on all forms of Multiplication, you will learn about this "X" or Duplex to solve problems like 45 x 45 :

45^2 = Duplexes of 4 / 45 / and 5

= 16 / 40 / 25

= 2,025

Vedic Maths stimulates the Pineal Gland (shown in the very centre of the brain) to know answers to mathematical problems instantly, intuitively and via pattern recognition. This is explained on the previous page as "VILOKANAM", the most important of the 16 Sutras that means "BY MERE OBSERVATION" !

EXTRA NOTES

ODE TO THE MATHEMATICIAN

港權，男童莊豐源獲准留港生活（上圖）。香港市

Beloved Teacher
You who have dedicated your Life
to the Teaching of Numbed Numbers
and deformed Formulae,
does it sadden you ?
each Day, each Year, each Decade
the repetition of classes repeating
that give a meagre and monotonous pay.
Beloved Teacher
what do you think of, late at night,
lying awake in bed, gazing at the ceiling,
planning your next lesson
with tired arms wrapped behind your head ?
Does it sadden you ?
that the bulk of the Children in your care
look away from their page
with a dull and distant, faraway stare.
Oddly, even the seeing or the hearing
of the very 4-syllabled word :
"Math - e - mat - ics"
baulks them, frightens them,
kineisiologically shuts them down
and turns them off ... pto

Communic8 with your kids

MATHEMATICS

Vectors, Tensors, Spinors

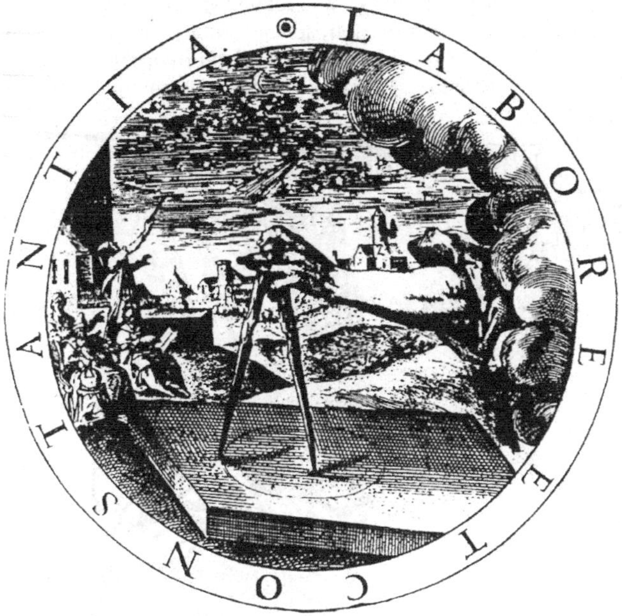

CONSTANT LABOR (Engraving, *A Collection of Emblems Ancient and Moderne* by George Wither, 1635) Wither writes, "Good Hopes, we best accomplish may, by laboring in a constant Way." Here the theme is repeated that the Work is accomplished by Working; in other words, the self-discipline of the Adept is a most important component of the Mystic Path. This message is repeated, in one form or another, through all the world's religious literature.

(ODE to the MATHEMATICIAN) cont...

Not a sound can be heard
as you stand there like a cop
except the ocassional sneaky spit
of a chewed up, masticated
fragment of a blue biro-top!
Beloved Teacher
you have illegally built your Shack of
Numbers without a D.A.
I invite you to tear this down
and dig through your pain,
through this cracked foundation, and
go much deeper
go beyond your fear of a loss of a job
go to the Core, to the Truth of Maths
to resurrect the ancient yogi's buried
Treasure Chest of flawless Vedic Diamonds.
The recrudescence of this priceless Knowledge
is the Light and the Wealth
upon which to build
the new Numerical University, with a
Higher Voltage that's conducive and right:
A highly sophisticated system of Intelligence
that takes the tears from mathematics
and imbues a sense of wonderment,
clarity, self-confidence and pto

HERMES (Illustration, *De Divinatione et magicis praestigiis* by Jean Jacques Boissard, ca. 1615) Hermes, the Greek derivative of Thoth, is also known as the wing-footed Messenger of the Gods, the God of Thieves, and the trickster. In the aspect of "Thrice-Greatest," or "Trismegistus," Hermes is closer to the majestic and intellectual Thoth. The Hermetic writings were compiled by different writers all using the same pseudonym.

(ODE to the Mathematician) cont...

mental agility
to all the Children in your care.
A Lightning Flash of neuro-transmitters
firing vedically in the brain is far superior
than the current dismal falling down of
digital rain.
If only you realized
how profound and inter-dimensional
your Algebra truly is;
I honour you, Learned One
in the highest sense of Maths Head or
Maths Master for ye are on the Cusp
of Self-Realizing that ye
are a Grand Teacher of the
ancient Star Language and
Frequency of Light:
A noble path indeed:
To potentially empower, awaken
and raise the Consciousness,
globally, of all the Children
in your care.
Your Number is up. pto

(Ode to the Mathematician) cont...

Beloved Mathematician,
now that you've been acknowledged,
glorified and heard the rave,
Tell me
can you continue to continue ?
which way are you going ?
Are you a Duck or

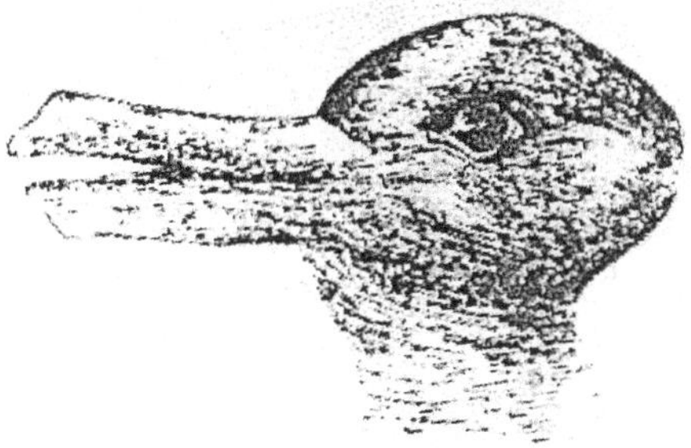

are you a Rabbit ?
Humbly yours
Jain 25/08/2001
I am at your Service.
It is time to be Creative
and take a Leap forward.
Leap and the Net appears!

It is Time
4 U 2 B
A Recipient

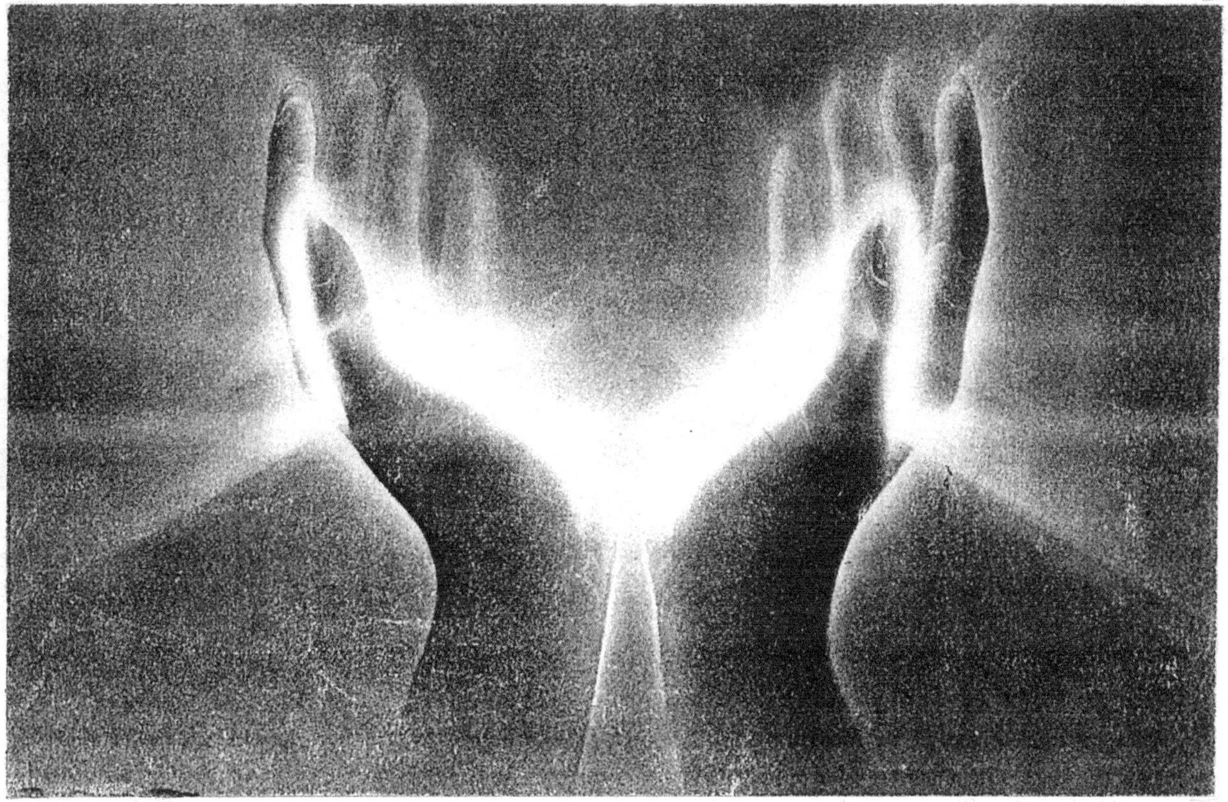

for the Truth
in
Vedic Mathematics.

the
PARABLE
of
DIVIDING
the
ELEPHANTS

PARABLE OF DIVIDING THE ELEPHANTS

(To elucidate the deeper meaning of the Vedic Sutra : "BY ONE MORE").

The Maharaja of an Indian state invited a travelling family of 3 wise Vedic Mathematicians (Grandfather, Father and Son) to teach and instruct his people plus his blind daughter, for a duration of one year, the art of numbers... As the day of departure arrived, he presented the 3 wise men with 17 Elephants, and with an expressed wish that the Grandfather gets $\frac{1}{2}$ of the Elephants, the Father gets $\frac{1}{3}$rd of that number and the Son gets $\frac{1}{9}$th!

The 3 Teachers decided not to leave until they had solved this puzzle, without any bloodshed. How could it be possible to divide the Elephants according to the expressed wishes of the Maharaja? Obviously, $\frac{1}{2}$ of the Elephants indicated clearly that the Grandfather will receive $8\frac{1}{2}$ Elephants! ... From a distance, the Maharaja and his blind daughter heard figures and fractions flying around...

"What use is an Elephant cut in two?" they argued, with hands flailing in the air. Biting their lips and scratching their heads, they remained in their triangle discussing the impossibility of sharing the elephants in whole numbers. "Maybe the Maharaja has erred, and we should let him know". So they tarried no longer to settle the affair and approached the Maharaja who was sipping tea with his daughter.

All five sat together, drinking tea. The Maharaja explained "Take your time, it can be done!"

Just then, a whistle was heard. The blind daughter had an "AHA" and said to her father: "Father, Father, I know how". She put 2 fingers in her mouth, made another quick loud whistle and an attendant soon arrived. "Bid me one Elephant from the stables, my dear Ahad"

An extra temple elephant walked its way to join the nearby herd of 17.

"Now," she said, "with my father's kind permission, I am temporarily gifting this elephant to you, my teachers, so now you have 18".

Without further explanation, the 3 teachers understood exactly what she was about to say.

"Here, Grandfather, you have $\frac{1}{2}$ of the 18" and she ordered 9 Elephants to walk his way.

"Here, Son of the Grandfather, you are entitled to $\frac{1}{3}$ of the "18" and she ordered 6 Elephants to walk his way.

With only 3 remaining in the original herd, her eyes met the boy, who she dearly loved, and will miss, saying:

"Your share is $\frac{1}{9}$ of the 18"

She gestured to Ahad and 2 Elephants walked towards the teenage boy.

There was only one Elephant remaining, and addressing all 3 men, she said: "I take back the temporary gift I had made". She gestured again to Ahad who sat upon its neck and rode it back to the stables.

The Maharaja was so pleased with his clever daughter and the 3 men were astounded.

She had learnt the true meaning of one of the main 16 Sutras: "BY ONE MORE" to solve a practical problem.

(INTERPRETATION)

• The Blind Girl is symbolic of the Feminine Brain, that realm of Intuition that utilized a Higher Sense to solve a riddle that the Logical Mind (symbolized by the 3 Men) was obstructed by.

• The Servant's name is "AHAD" which means "ONE"

• "BY ONE MORE THAN THE PENULTIMATE DIGIT" Sutra effortlessly and brilliantly solve mathematical problems like :

1) 25^2. 5 is the last digit which is squared = _/<u>25</u> and the penultimate (literally means "just before the last") digit is 2. "By" means "x" $\therefore$ 'By One More' means 2x3 $\therefore$ $25^2 = 2 \times 3/25 = 6/25$.

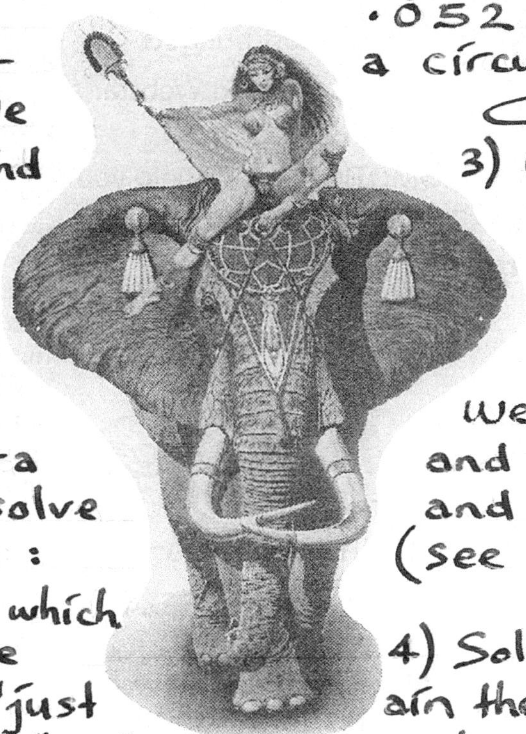

2) $\frac{1}{19}$. By One More means : we multiply (Ekadhika) by one more than the 1 of $\frac{1}{19}$ = x 2 to convert $\frac{1}{19}$ into a decimal = miraculously : $\cdot\dot{0}52631578/94736842\dot{1}$ a circulating 18 digit answer !

3) What is the 20th Triangular Number ?

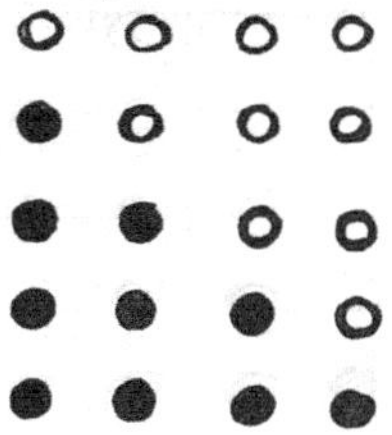

By understanding the 4th Δ No. = 10 $= \frac{1}{2}(4 \cdot \times 5) \longrightarrow$ We apply "BY ONE MORE" and say $\frac{1}{2}(20 \times 21)$ and get our answer = 210 (see Page 41 for more details).

4) Solving SQUARE ROOTS to ascertain the correct choice of 2 possible numbers for the unit's place eg: $\sqrt{4096}$ Must be a 2 digit answer (40 / 96) _/_. Last digit is a 4 or 6, but which? First digit is 6 (as $6^2 < 40$). Say "By One More" $= 6 \times 7 = 42$. Since $42 > 40$ we choose the 4 not the 6 $\therefore \sqrt{4096} = 64$!

(This Parable, from the previous pages, has been adapted and changed from Shakuntala Devi's book "FIGURING: THE JOY OF NUMBERS"

KNOWLEDGE/IGNORANCE, letter regrouping, by Jeffrey Scott, Los Angeles, Calif.

Dividing the Elephants

The Maharaja of an Indian State invited three artists to perform at his court. He was much impressed, and honoured them with gold and gifts.

On the day of their departure he presented the three of them with seventeen elephants, expressing a wish that the oldest of the

artists should get one-half of the number of elephants, the middle one one-third of the number, and the last one one-ninth.

When the artists returned home, they were absolutely puzzled: how could they divide the elephants according to the wishes of the Maharaja? One-half of the seventeen elephants meant that the oldest artist gets eight and a half elephants. What good is an elephant cut in two! They argued, and the leading citizens of the town were called to settle the dispute. They tried in vain to carry out the will of the Maharaja for they found themselves constantly held up, by the impossibility of making any share a whole number of elephants. They were about to give up in despair when someone in the crowd suggested that the temple priest, a wise, very learned man, be called in to settle the affair.

The temple priest listened to the whole story in detail, got up from his seat and asked to be excused. Then he went back to the temple and returned in a few minutes with the temple elephant. 'Now,' he remarked, 'the problem can be easily solved. I am temporarily gifting this temple elephant, and pooling it in with the other elephants.'

There were altogether eighteen elephants, with the temple elephant thrown in.

The priest addressed the oldest artist. 'Here, you take away half the number of elephants,' he said and presented him with nine elephants.

Then he looked at the middle artist and remarked, 'You are entitled to one-third of the number, and here are your six elephants.' He gave away six elephants to him.

Now there were three elephants remaining including the temple elephant.

Then to the youngest artist he said, 'Your share is one-ninth of the total number of elephants. Therefore you get two elephants.' He gave away two elephants to him.

The remaining elephant was the temple elephant. 'I take back

the temporary gift I had made.' So saying he walked away with his elephant, back to the temple. The will of the Maharaja was followed to the letter.

Mathematical genius

Shakuntala Devi is an Indian woman of immense arithmetical ability who demonstrates her amazing skills in public. It took just a second for her to find that the cube root of 332,812,557 is 693. In the 1970s she calculated the 23rd root of a 201-digit number. The problem had been devised by Texan students. A computer took one minute and 13,000 instructions to check her answer and prove her right.

Shakuntala Devi

"All along I have cherished a desire to show those who think mathematics boring and dull just how beautiful it can be. This book is what I offer."
Shakuntala Devi

Creation of the universe by Lord Brahmā, the great grandfather of all living beings.

HISTORY OF VEDIC MATHEMATICS from "BHAGAVAD - GĪTĀ"

The Bhagavad Gita (literally: "the Song of God") is one of the oldest and most popular book in existence.

In Chapter 9.2, it says:

"This knowledge is the king of education, the most secret of all secrets. It is the purest knowledge, and because it gives direct perception of the self by realization, it is the perfection of religion. It is everlasting, and it is joyfully performed".

These words were spoken by Kṛṣṇa to Arjuna, on the battlefield.

The impact of this, plus the following quotes, places Vedic Mathematics into a category of Royal Knowledge that is transcendental and the king of all confidential knowledge. It is a universal language that can be executed without difficulty. pto

Painting representing the universal form of the Goddess of Speech, by which the named world is brought into existence. In her body are saints, gods and kings. Rajasthan, 19th century. Gouache on paper 11 × 8 in. Ajit Mookerjee, New Delhi.

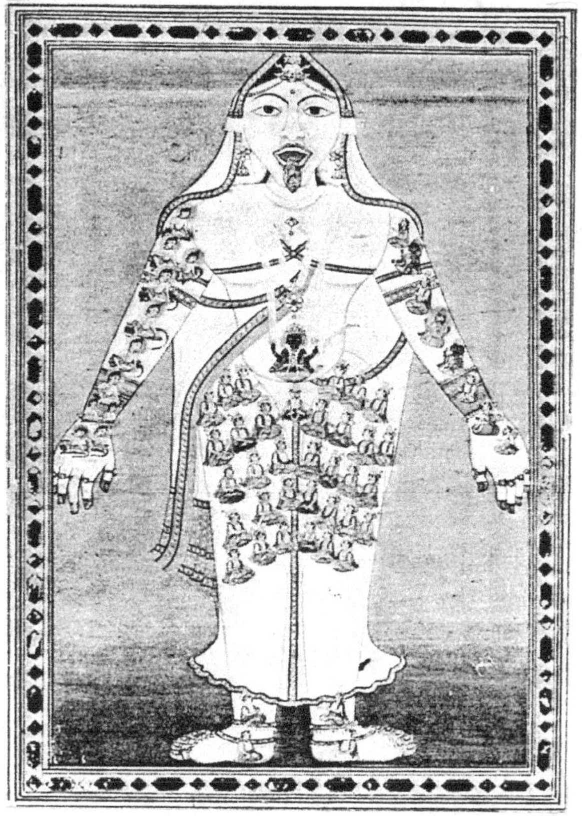

Therefore, it can easily and joyfully be adopted to build the mental muscle.

The origin of the Vedic Knowledge is explained in the beginning of Chapter 4.1

"The Personality of Godhead, Lord Sri Kṛṣṇa said: I instructed this imperishable science of yoga to the sun-god, Vivasvān, and Vivasān instructed it to Manu, the father of mankind, and Manu in turn instructed it to Ikṣvāku".

Thus the history of Vedic Knowledge came from the Stars and was delivered to the royal order of all planets, beginning from the sun planet. The king's role was to impart this knowledge to protect their citizens from material bondage or lust. Human life is meant for cultivation of spiritual knowledge and the executive heads of all states and all planets are obliged to impart this lesson to the citizens by right education, culture and devotion, in order to pursue a successful path in Life.

In this millenium, the Sun-God is known as VIVASĀN, the King of the Sun, which is the origin of all planets within the solar system. The sun represents the Eye of the Lord and supplies heat and light. Kṛṣṇa originally made Vivasvan His first pto

THE SCIENCE OF SELF REALIZATION

इयानसावीश्वरविग्रहस्य
यः सन्निवेशः कथितो मया ते ।
सन्धार्यतेऽस्मिन् वपुषि स्थविष्ठे
मनः स्वबुद्ध्या न यतोऽस्ति किञ्चित् ॥३८॥

"The Cosmic Manifestation"

disciple to understand the science of the Vedic Knowledge like in the Bhagavad-Gita which is a standard book of knowledge coming down from time immemorial.

In fact, the history of this Vedic Knowledge can be traced back to specific Ages, and is described in the Mahabharata eg in Treta-Yuga (which lasts 1,200,000 years) Vivasvān delivered the Vedic Codes to Manu. Manu, the father of mankind (like the Christian's Adam) gave it to his son Maharaja Iksvaku, the king of this Earth planet and forefather of the Raghu dynasty, in which Lord Ramacandra appeared.

Then the Dvapara-Yuga appeared and lasted for 800,000 years.

At the present moment, we have just passed through 5,000 years of the Kali-Yuga which will last 432,000 years. But this is an Age of Darkness, of forgetfulness of our own Light. Thus 5,000 years ago, we lost sight of this Vedic transmission and a clumsy, corrupted form of written mathematics was necessarily adopted. Prior to this, the ancients learnt

pto

mathematics as an oral transmission, and as Kali Yuga set in, due to forgetfulness, the Sutras were then written down and evolved to the form used today. Somewhere around 700 A.D. in the Caliph's Court (Iraq) in Bhagdad, the original knowledge from the Aryan Race near the Hunza Valley, had got lost or ceased to be exchanged during the Islamic Wars. If we continue to do maths on calculators the global brain will atrophy. That is why mental calculation, in the Vedic style, is a high science and is paramount in reclaiming the genius of our original memory and knowledge.

In fact, the Bhagavad Gita was spoken at least 120,400,000 years ago (ie the Lord Kṛṣna speaking to the Sun-God Vivasvān) but has only been extant for 2 million years, in human society. In sanskrit, this Vedic Knowledge is termed "APAURUSEYA" or Super Human.

It is important that the revival of this saintly knowledge be reclaimed by the People as already N.A.S.A are using this for 'Artificial Intelligence', and yet, ironically, the people of Vedic Math's Motherland, India, are struggling to get it re-introduced into their school curricula!

pto.

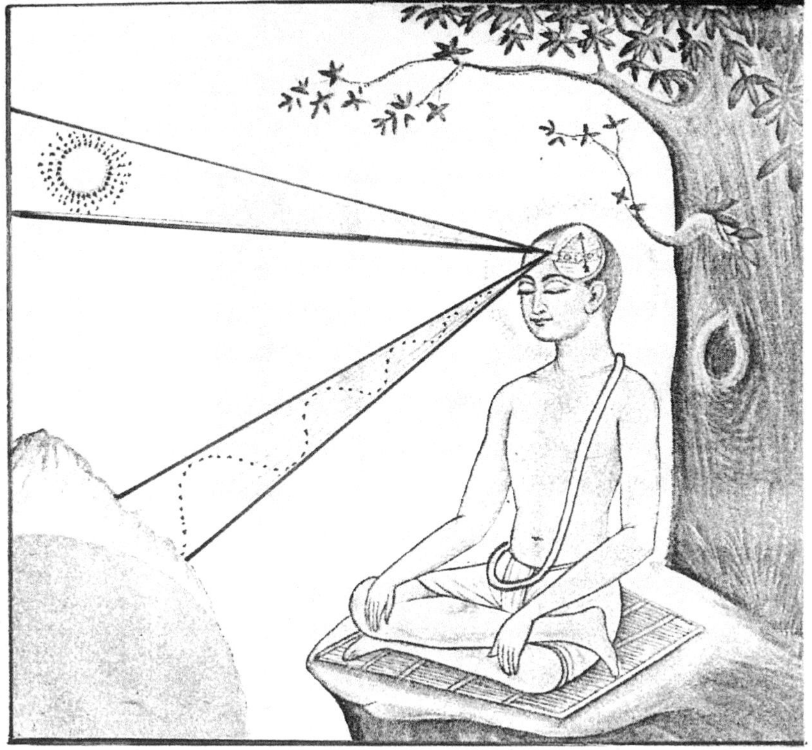

There are individuals and some professors egotistically claiming this Vedic Maths as their own system without acknowledging the Vedic Source. Some individuals are claiming occult powers. This is all ego-based and misguided. To teach Vedic Mathematics in the current Western system without appropriate meditation upon the deeper meaning of the 16-Sutras, is a waste of time.

In the Bhagavad Gita, Chapter 4·2, it says:

"This supreme science was thus received through the chain of disciplic succession, and the saintly kings understood it in that way. But in course of time, the succession was broken and therefore the science as it is appears to be lost".

To conclude, the following is the definition of "VEDA" given by Sri Bharati Krsna Tirthaji:

"The very word 'VEDA' has this derivational meaning ie the fountainhead and illimitable store-house of all knowledge. This derivation, in effect, means, connotes and implies that the Vedas should contain within themselves all the knowledge needed by mankind ... spiritual and worldly ... for the achievement of all-round, complete and perfect success"...

Self-Realization Fellowship
Founded in 1920 by Paramahansa Yogananda, author of the spiritual classic *Autobiography of a Yogi*

SUPER-SCIENCE IN ANCIENT INDIA

India may have had a superior civilisation with possible contacts with extraterrestrial visitors, and the flying devices called 'Vimanas' described in ancient Indian texts may underline their possible connections with today's aerospace technology, an Italian scientist told the World Space Conference here today.

Dr Roberto Pinotti asked the delegates to examine in detail the Hindu texts instead of dismissing 'all the Vimana descriptions and traditions as mere myth'.

"The importance of such studies and investigations could prove to be shocking for today's man because the existence of flying devices beyond mythology can only be explained with a forgotten superior civilisation on earth," he said.

According to Dr Pinotti, the huge 'Shakuna Vimana' described in the text 'might be defined as a cross between a plane and a rocket of our times and its design might remind one of today's space shuttle'.

'Surely, it expresses the most complex and sophisticated aeronautical design among all the other descriptions of Vimanas mentioned in the *Vymaanika-Shaastra* he said. He described the author of the treatise *Vymaanika-Shaastra* as a man 'attempting to explain an advanced technology'.

weapons and the fact that Vimanas resembled modern unidentified flying objects would suggest that India had a "superior but forgotten civilisation."

"In the light of this, we think it will be better to examine the Hindu texts and subject the descriptive models of Vimanas to more scientific scrutiny", he said.

Pointing out that Indian Gods and heroes fought in the skies using piloted vehicles with terrible weapons, Dr Pinotti said they were similar to modern jet propelled flying machines.

He said certain descriptions of the Vimanas seemed 'too detailed and technical in nature to be labelled as myth'. He cited various texts to show there were 32 secrets relating to the operation of Vimanas, some of which could be compared to modern-day use of radar, solar energy and photography.

Quoting from *Vymaanika-Shaastra* he said the ancient flying devices of India were made from special heat absorbing metals named 'Somaka, Soundalike and Mourthwika'.

He said the text also discussed the seven kinds of mirror and lenses installed aboard for defensive and offensive uses. The so-called 'Pinjula Mirror' offered a sort of 'visual shield' preventing the pilots from being blinded by 'evil rays' and the weapon 'Marika' used to shoot enemy aircraft 'does not seem too different from what we today called laser technology,' he said.

According to the Italian expert, the 'principles of propulsion as far as the descriptions were concerned, might be defined as electrical and chemical but solar energy was also involved.

For instance, the 'Tripura Vimana' mentioned in *Vymaanika-Shaastra* was a large craft operated by 'motive power generated by solar rays,' Dr. Pinotti said.

Dr. Pinotti, who has made an exhaustive study of the history of Indian astronautics, said another text, *Samarangana Sutradhara* had 230 stanzas devoted to the principles of building Vimanas and their use in peace and war.

He said ancient Aryans knew the use of the element 'fire' as could be seen from their 'Astra' weapons that included Soposamhara (flame belching missile), Prasvapna (which caused sleep) and four kinds of Agni Astras that travelled in sheets of flame and produced thunder.

He said the car that was supposed to go up to Suryamandal (solar system) and the Naksatramandala (stellar system) cannot be dismissed as a myth because of the 'technical nature' of its description.

NEXUS•49

FEBRUARY-MARCH '93

(Source: The Hindu, *12 October 1988, Bangalore, India)*

RUKMA VIMANA

by David Hatcher Childress, author of *Vimana Aircraft of Ancient India & Atlantis.*

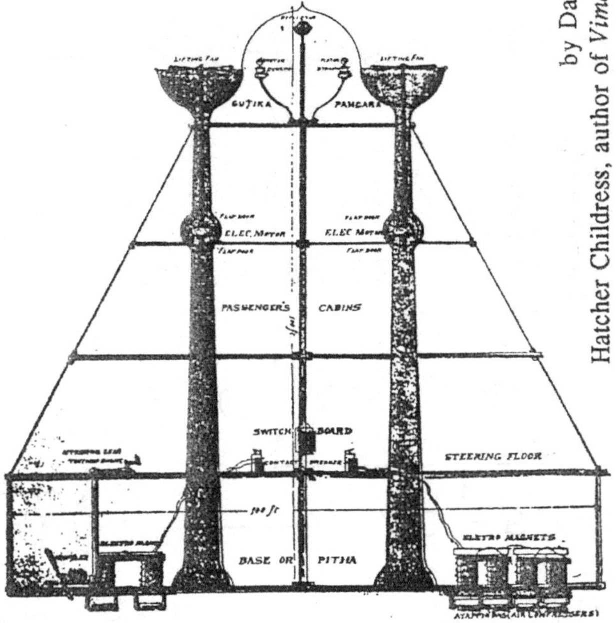

A 1923 drawing by T.K. Ellappa of Bangalore, India of a Vimana (Vimaana) prepared under instruction from Pandit Subbaraya Sastry of Anckal, Bangalore.

on the ORIGIN of ZERO

Perhaps the greatest discovery of all time was the Invention of ZERO. Without it we could not have sent rockets to the moon nor created silicon-chip technology.

The very concept of Zero is attributed to the Hindus, not the Arabs as falsely believed today. The Hindus needed a symbol that gave positional value in their base 10 number system. Originally, it was indicated by a small circle which they called "SUNYA", a Sanskrit word meaning "Vacant, void". The Arabs, circa 800 AD translated this to "SIFR" and then subsequent changes were made from the Latin to the French to the German then to our current English. (The Origin of Zero in the **Vedas** is taken from "Mathematics As Known To The Vedic Samhitās", by M.D. Pandit, 1993).

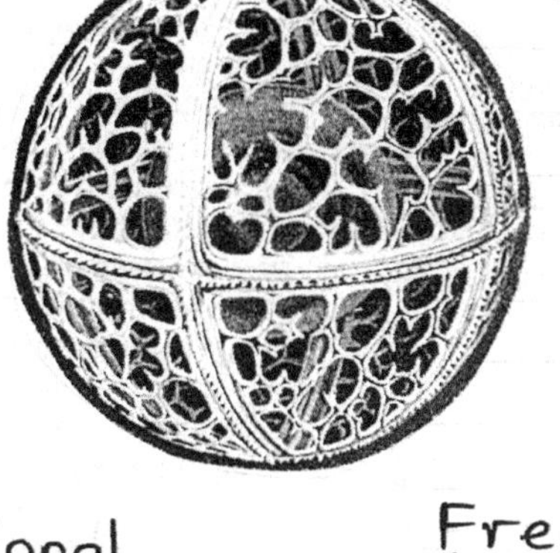

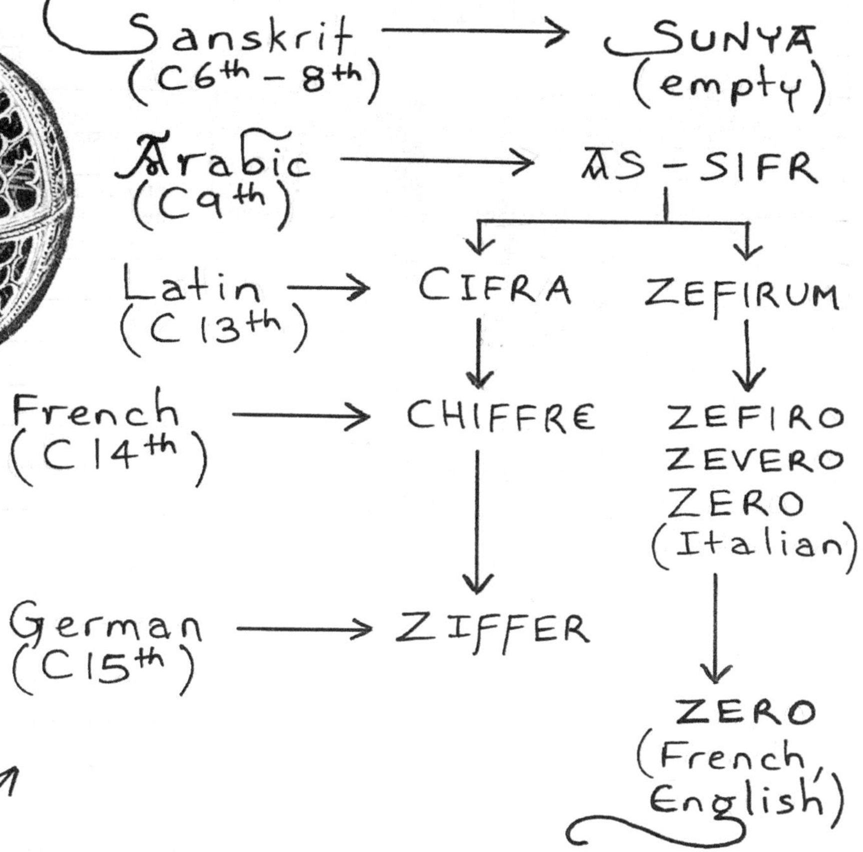

[The spherical graphic above is called "Radiolaria", the skeletal remains of a marine Protozoa].

JAIN'S ULTIMATE MAGIC SQUARE:

- The Sums of the 3 Columns = 0
- " " " " 3 Rows = 0
- " " " " 2 Diagonals = 0

Thus The CONSTANT equals ZERO

THE UNIVERSAL ONE

When man knows this language he will then think in it.

When he thinks in light he will then know that he is the Son of God, that he is One with God.

When man thinks in light he will then know that he is universal Mind; that he is all that is; that he is thinking Mind expressing his thinking in light; that he is omnipotent, omniscient and omnipresent.

He will then know that he is life and that life is eternal and that there is no death; and that he has neither place nor position in space or time; and that dimension is as though it were not; and that there is but One, and that One He, and you, and me, and the rolling hills and the sands of the sea, and the clouds and stars and the little violet blooming in the meadow.

Already man has gone a far way toward the solving of the great secret of this creating universe which lies hidden in the elements.

Within God's alphabet of light the locked secret has long awaited the master key of knowledge to open the doors to man.

Man is Mind.

Mind creates and controls the energies of the elements.

Mind may rearrange these energies and combine them in accord with its desire.

(from Walter Russell's book).

THE FORMULA OF THE LOCKED POTENTIALS

4+3+2+1+0=1-2-3-4

CONCEPTS of ZERO

• We tend to imagine that our counting system begins at zero and goes on from there indefinitely: we learn later that negative numbers are as valid and real as our 'positive' numbers. Zero thus manifests as the centre of an intersection of two fields of numbers, one positive and one negative.

• A negative number may be viewed as a 'monetary debt'.

• In regards to time, Zero is the imagined boundary between past and future, between negative and positive time.

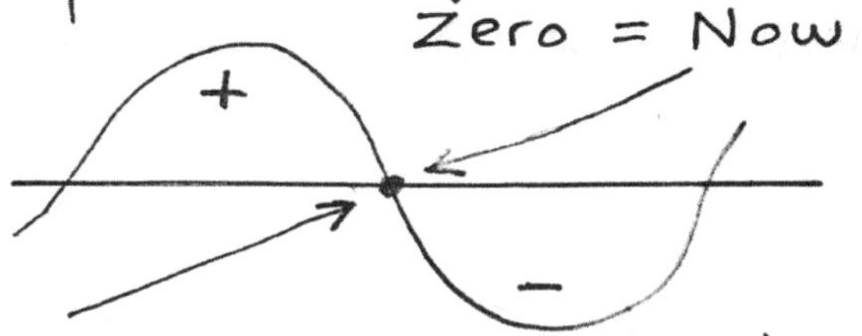

Zero is also Einstein's uncurved space : the moment of NOW in time.

• Zero is also symbolic of the birth of Christ on Earth in regards to Jesus' birth setting Earth time calendarically back to Zero. In Barbara Hand Clow's book (1995) "Pleiadian Agenda", she says: "Christ manifested at ZERO POINT in history to evolve humanity into empathy, the highest vibration of the Age of Pisces (♓). Empathy opens humans to spiritual access, and Christ came as a model of the 9-Dimensional Human, which is what you will all become during the Age of Aquarius

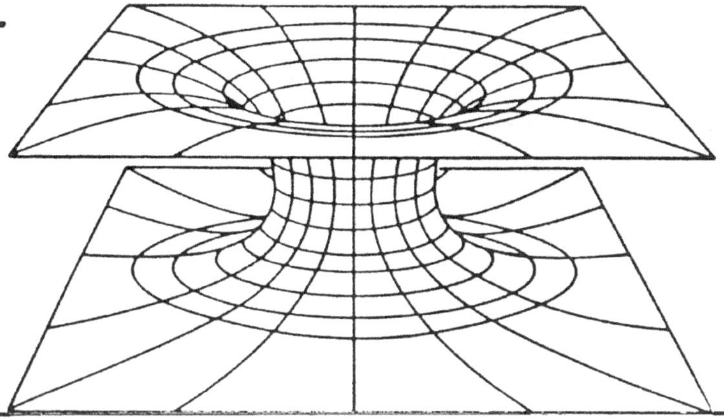

FATHER/MOTHER SPIN A WEB

(ON THE NATURE OF ZERO):

(The following creation of the DECAGON (a 10-sided Polygon with all sides and radii in The Divine Proportion) is adapted from "The Secret Doctrine", by Madame Blavatsky.

The Circle / Zero is the symbol for CosmoGenesis. It is Sunya, The Void, The Cosmos. The one circle is a Divine Form from which all proceeds, whither all returns. A description of the following four circular forms follows:

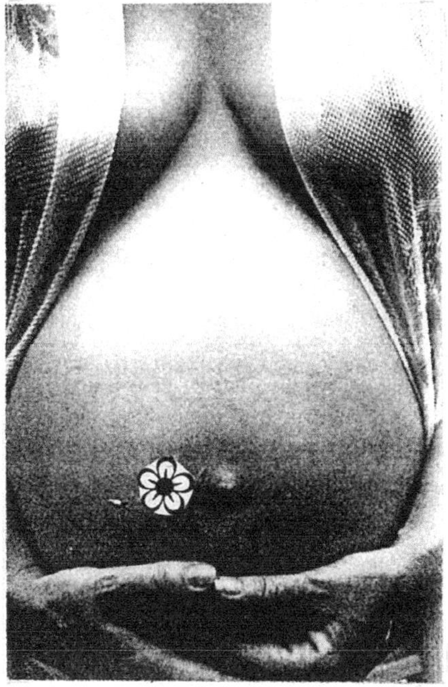

1) The plain disk represents Zero, the Boundless, ever-eternal nature, sexless, the Fool in The Tarot, infinite "ALL THAT IS".

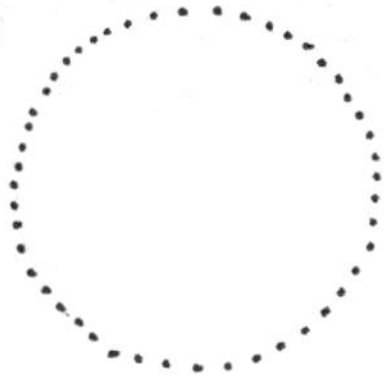

2) The point within the circle now represents the First Logos, Brahman-Pradhana, the point indicating differentiation or potential Space within abstract Space

3) In its 3rd stage, the point is transformed into a diameter. (Horizontal lines are considered Feminine, and vertical lines masculine). This Feminine, Mother-Nature quasi-manifestation represents the Second Logos, Brahma-Prakriti or Svabhavat.

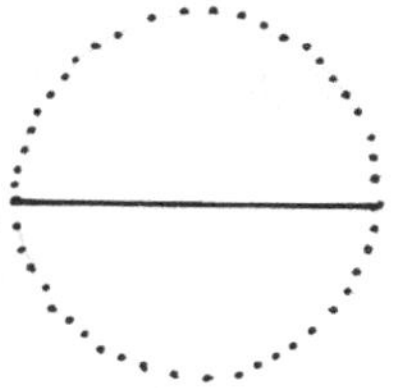

4) When the Vertical line crosses the diameter, the Third Logos, Mahat, is represented, but it is not static, its extended arrow represents motion, or the action of manifestation. Father-Mother begin to spin a web, THIS IS THE ORIGIN OF THE MUNDANE CROSS.

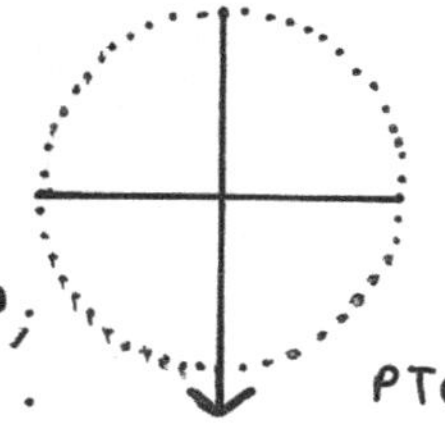

PTO

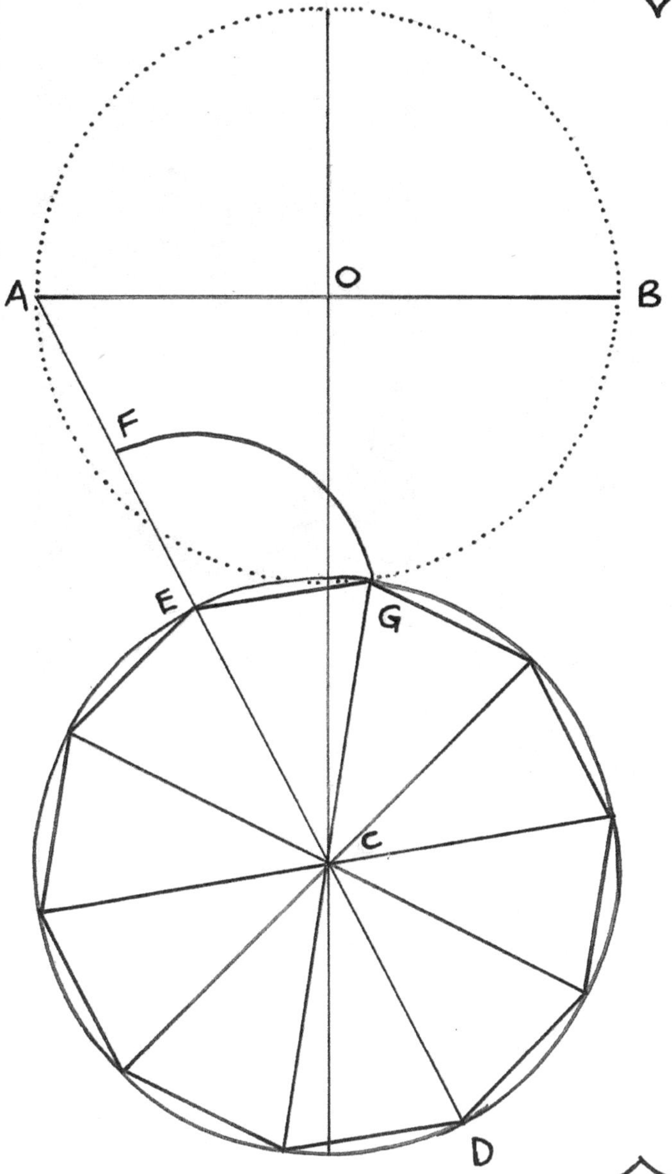

On p29 of 'the Secret Doctrine', Verse 10, Stanza 3 of the Book of Dzyan :

"Father-Mother spin a web whose upper end is fastened to Spirit - the light of the one darkness - and the lower end to its shadowy end, matter, and this web is the universe spun out of the 2 substances made in one, which is Svabharat."

Its like God-Goddess, wanting to experience Him/Her Self replicated HimHerSelf. Man-Woman was made in the Image of God-Goddess. Thus "O" became "8" ie The 1 became The 2.

Notice here in our diagram that the circle of Spirit has descended a distance of its own diameter AB. The lower circle of matter occupies a new position in space, like a fertile seed.

The diameter AB has become a Vector, passes through the lower circle's centre "C" to form AD. Point 'A' was our original fixed point, now point 'D' has been created, and point 'E' to begin the creation of the Golden Mean Decagon whose measurements are akin to the human-body, planetary distances, flowers and D.N.A.

The lengthened line AD is cut by the circle at the point E, and a remarkable relationship is now observed between the line segments: It is traditionally expressed as : The original line AD, known as the 1st, is to the longer segment DE, known as the 2nd, is pto

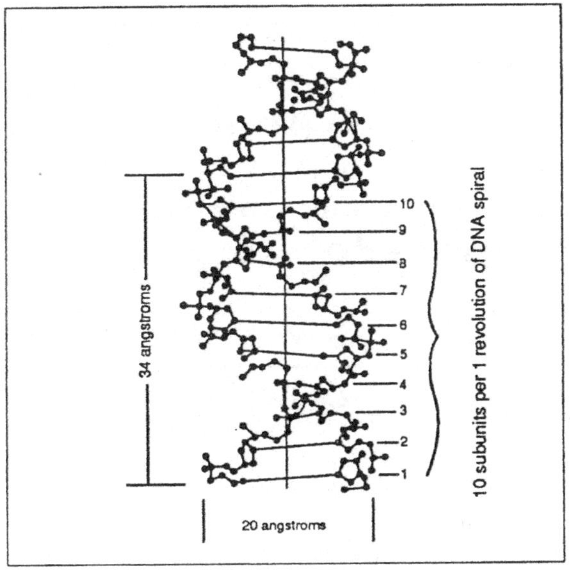

One 360 degree turn of DNA measures 34 angstroms in the direction of the axis. The width of the molecule is 20 angstroms, to the nearest angstrom. These lengths, 34:20, are in the ratio of the golden mean, within thelimits of the accuracy of the measurements. Each DNA strand contains periodically recurring phosphate and sugar subunits. There are 10 such phosphate-sugar groups in each full 360 degree revolution of the DNA spiral. Thus the amount of rotation of each of these subunits around the DNA cylinder is 360 degrees divided by 10, or 36 degrees. This is exactly half the pentagon rotation, showing a close relation of the DNA sub-unit to the golden mean.

in the same proportion as is the longer (ED) is to the shorter segment AE, known as The 3rd. This relationship $\frac{AD}{DE} : \frac{DE}{AE}$ is called the Golden Measure, the Divine Proportion or the Phi Ratio whose symbol is "ϕ" the ZERO penetrated by the ONE. Combining the firST to the secoND to the thiRD gave the origin of our universal ST-ND-RD of measurement.

Often, in Sacred Geometry, we let the radius of the original circle AO to be 1 unit. Thus the diameter AB =2 units, which means "OC" also =2 units. Regarding the right-angled triangle AOC, length of AC = $\sqrt{1^2+2^2} = \sqrt{5}$ or 2·236. This $\sqrt{5}$ is a part-expression of Phi ϕ.

Cutting AC at the point E reduces AC by the length of the radius, so that AE = 2·236 − 1 or 1·236. ED = 2 and the line AD = 1·236 + 2 = 3·236. Thus:

1·236 : 2 = 2 : 3·236 or dividing conveniently by 2

·618 : 1 = 1 : 1·618 = Phi = ϕ

Now, similar to dividing by 2, we can achieve the same geometrically by taking the midpoint of AE to get F and taking the ratio of FE and the radius EC to form the first side of the long-awaited for DECAGON. EF : EC = ·618 : 1 = EG.

The toroidal, or donut-shaped, wave fields that surround the heart when it fires. These fields are *fractal*—fields within fields radiating out from the heart.

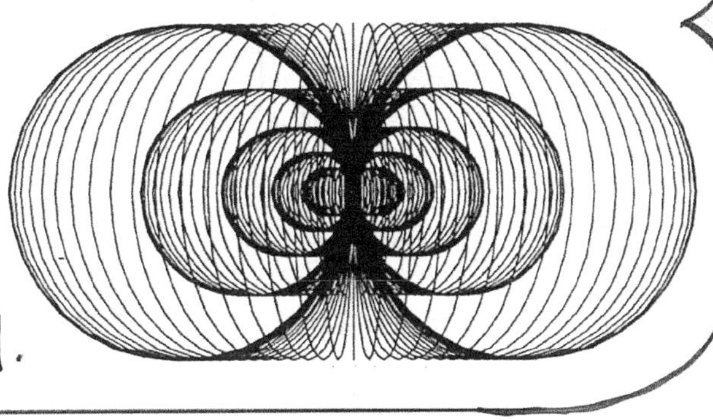

[This is an excerpt from "THE BOOK OF PHI" by JAIN, to be published and released in 2002, after 20 yrs research].

1) Solve $9{,}997^2$?
"By whatever the deficiency, lessen it by that much and set up the square thereof."
Use Base 10,000.

```
= 9997 - 0003
  9997 - 0003
  ------------
  9994 / 0009
```

2) Divide 202 by 99 ?
Using base 100, a divisor of 99 has an Operator of 01. We write this 01 underneath the 99 and set out like this:

```
99 ) 2 | 0 2
01     |
     ↓ | 0 2
   ----+-----
     2 | 0 4
```

The 2 digit operator (01) determines the vertical line to be inset 2 digits from the right. Multiply 2 × 01 and add. Answer = 2 rem 04.

12

GOOD EXAMPLES DEMONSTRATING THE INVENTION OF ZERO

perhaps the greatest feat of all TIME

3) Divide 11,111,111 by 9,989

```
9989 ) 1 1 1 1 | 1 1 1 1
0011   | 0 0 1 | 1
       | | 0 0 | 1 1
       | | | 0 | 0 1 1
       ↓ ↓ ↓ ↓ | 0 0 2 2
       --------+---------
       1 1 1 2 | r3 3 4 3
```

nb: The Operator (0011) is found by saying: "All from 9 and the Last from 10", regarding 9989. This is the definition of Nikhilam Sutra.

4) Multiply 397 × 303.
Using "By One More" Sutra:
We notice that both numbers begin with a 3
= 3/97 × 3/03 and that
(97) + (03) = 100.
We must have 4 place units on the Right Hand side 100 × 100.
= 3×4 / 97 × 03
= 12 / 2̲9̲1̲ = 120,291

INVENTION of ZERO: EXAMPLES

5) Divide 27 by 5.
Use the Sutra:
"TRANSPOSE AND APPLY"
$= 27 \times 2 \div 5 \times 2$ by multiplying both sides by 2.
$= 54 \div 10$
$= 5 \cdot 4$

6) Solve $(1004)^3$ by use of Yavadunam.
When Cubing, we set up 3 parts to the answer.

_____ / _ _ _ / _ _ _

- add double the surplus $= 1004 + (2 \times 4) = 1012$ using Base 1,000.
- Multiply the original excess of 4 by the new excess of 12
- Cube the surplus or excess $= 4^3 = 64$.

Combining,
$1004^3 = 1012 / \_48 / \_64$
but we are in Base 1,000 and require 3 digit segments
$= 1012/048/064 = 1{,}012{,}048{,}064$.

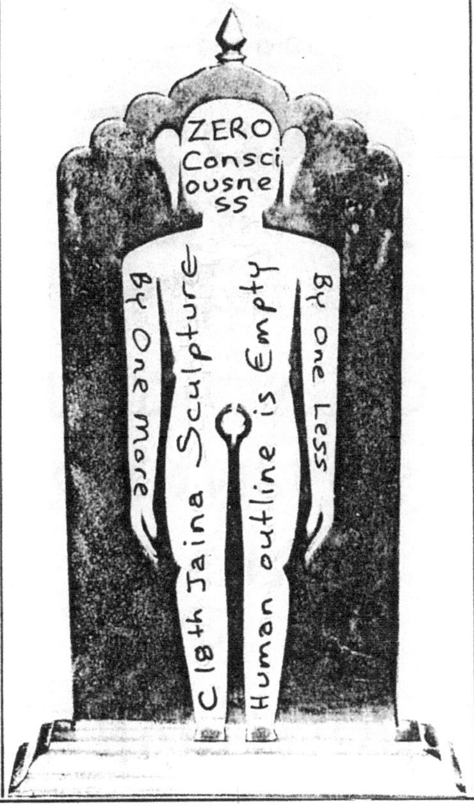

7) Solve 99×7.
must write it as: 99×07.
Say "By One Less" than $07 = 06$ and subtract (07) from 100
$= (07) - 1 / 100 - (07)$
$= 06/93$ or 693

8) Solve 999×65
Must write it as 999×065
Say, "By One Less" than 065 and subtract (065) from 1000.
$= (065) - 1 / 1000 - 065$
$= 064 / 935$ or 64,935.
nb: the quick way to subtract a number from 1000 is to use Nikhilam Sutra and say:
"All from 9 and the Last from 10"
thus $1000 - 065$
$= (9-0)/(9-6)/(10-5)$
$= 935$.

9) Solve $99{,}999 \times 798$
Write as 99999×00798
Say 'By One Less' than 00798 and subtract 00798 from 10,000 $= 00797 / 99202$ using "All from 9, Last from 10".

10) What is the $\sqrt{\cdot 000049}$?
Pair off all the digits
= ·00 / 00 / 49
Take the Square Root thereof.
= ·007

11) What is 999 x 7 ?
must be written as
999 × 007 as we are
in Base 1,000.
Say, 'By One Less' than (007)
and subtract (007) from 1000.
= (007) −1 / 1000 − (007)
= 006 / 993
= 6993

12) Divide 120,160 by 993
= 993) 1 2 0 | 1 6 0
007

1	2	0	1	6	0
↓	0	0	7		
↓	↓	0	0	14	
↓	↓	↓	0	0	0
1	2	0	8	20	0
			10	0	0
			−9	9	3
1	2	1	rem 0	0	7

INVENTION of ZERO EXAMPLES DEDICATED to JAMES BOND

"I suspect
that
James Bond
Agent 007
is a
reincarnation
of a great
Hymalayan
Vedic
Mathematician."

the SANSKRIT NAMES for the POWERS of TEN

10^0	=	1	eka
10^1	=	10	daśā
10^2	=	100	śata
10^3	=	1000	sahasra
10^4	=		ayutam
10^5	=		niyutam or lakṣa
10^6	=		prayutam
10^7	=		arbuda or koti
10^8	=		nyarbudam or abja
10^9	=		kharva
10^{10}	=		nikharva
10^{11}	=		mahāpadma
10^{12}	=		śaṁku
10^{13}	=		samudra
10^{14}	=		madhya
10^{15}	=		anta
10^{16}	=		parādha

(taken from "Mathematics as Known to the Vedic Saṁhitās").

HOW to COUNT from 1 → 22

1	eka
2	dvi
3	tri
4	catur
5	pañchan
6	ṣaṣ
7	saptan
8	aṣṭan
9	navan
10	daśan
11	ekadaśan
12	dvadaśan
13	trayodaśan
14	caturdaśan
15	pañcadaśan
16	sodaśan
17	sapadaśan
18	astādaśan
19	navadaśan
20	vimśati
21	ekavimśati
22	dvāvimśati

ZERO

Here is an extrapolation of the 9x9 Vedic Square in its 10x10 form. Shown below is Leonardo da Vinci's exploration of this timeless table.

The diagram on the top right shows all the Last Digits (Adyam, in Sanskrit) that end in Zero, being circled. The diagram below that shows "ALL THE ZEROES" connected to form the stunning Octahedral Crystal that Magician's step into for auric and Spiritual Protection, Centredness, + Strength.

Exploring the Last Digits, ending in Zero of the 10 x 10 Natural Square:

Formation of the 3-Dimensional DIAMOND CRYSTAL.

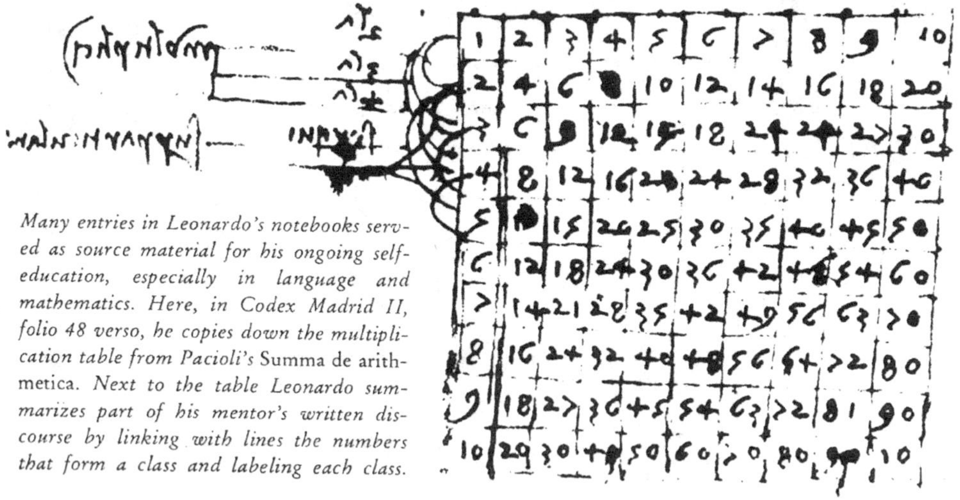

Many entries in Leonardo's notebooks served as source material for his ongoing self-education, especially in language and mathematics. Here, in Codex Madrid II, folio 48 verso, he copies down the multiplication table from Pacioli's Summa de arithmetica. *Next to the table Leonardo summarizes part of his mentor's written discourse by linking with lines the numbers that form a class and labeling each class.*

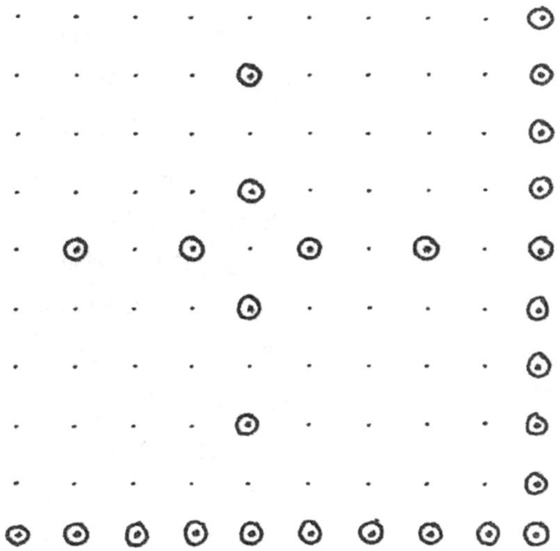

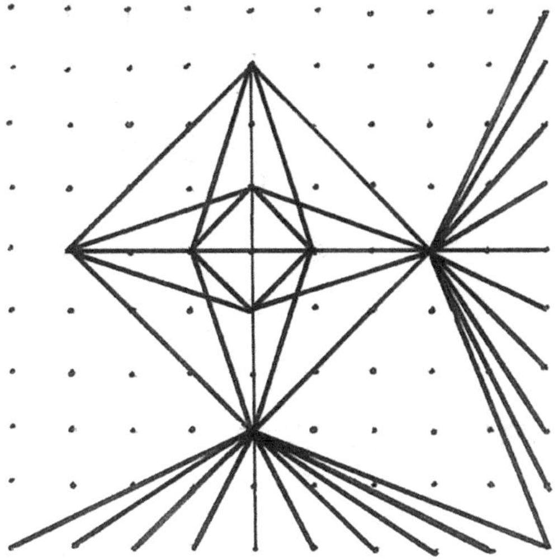

ACKNOWLEDGEMENTS

FOR GRAPHICS USED IN "EXTRA NOTES"

P116 : Photo from the front cover of "Hands Of Light" by Barbara A. Brennan.

P123 : Text and photo of Shakuntala Devi on Left-Hand-Side is copied from "How Maths Works" by C. Vorderman 1996.

P124 : Hare Krishna books, translated by Prabhupada.

P126 : ibid

P127 : ibid (top right): "As SuperSoul, the Supreme is situated in everyone's heart, also within every atom". Bottom Right : The Divine Eye or Eye of Wisdom of a meditator from "Atma Vijnana" or "Science of the Soul" by Vyas Devi Ji. Artist unknown/unacknowledged.

P129 : Article rearranged from "Nexus" magazine 1993, an Australian watch-dog on alternative medicine + energy.

P130 : Skeletal Sphere from the remains of a single-celled protozoa 'Radiolaria' taken from 'Art Forms In Nature' by Haeckel.

P132 : An extraterrestrial Vortice of a rotating black hole where 2 worlds are connected with each other. From 'Vortex Flow in Nature and Technology' by Hans. J. Lugt

Bottom Right : Logo on 'Vibrational Healing' by Darpan, Byron Bay.

P133 : Byron Bay Journal, 2001, called "Belly" by Sarah Nicholson. Photo by Elk Anstey.

P134 : Decahedral geometry from 'The Mathematics of The Cosmic Mind'. L.G. Plummer.

P135 : Both graphics from Dan Winter's banned book : 'Alphabet of The Heart'.

P137 : C18th Jaina Sculpture. Body Shape is Void !

P140 : Bottom Left. Multiplication Table + text reprinted from "Leonardo The Scientist".

P142 : The Work of Callum Coats.

hum sah
so hum
om

HAVE YOU EVER WATCHED A NEW FERN LEAF UNSPIRAL ?

"COPY AND COMPREHEND NATURE" (Callum Coats)

"ETHER" from the Greek word 'Aither' meaning : "Upper Air, Sky" The upper regions of Space, clear sky, the Heavens. The ancients all-pervading medium transmitting Light + ALL-KNOWINGNESS

How can a torn leaf grow back into its original form? How can an Amputee 'feel' the missing limb? How can a lizard or salamander grow back its missing tail? Easy. Because its still there, etherically!

In like manner, "How can a young Vedic Mathematician just "KNOW" the final answer? Easy. Because we are all, by birth, naturally brilliant and originally solved all calculations mentally "BY MERE INSPECTION" (one of the 16 Sutras) or Intuitiveness. The Etheric Answer is still there, elegantly disguised in the Question's visible PATTERN. You see "THE SHAPE" and merely fill in the Digit Answers. Life Was Meant To Be Simple.

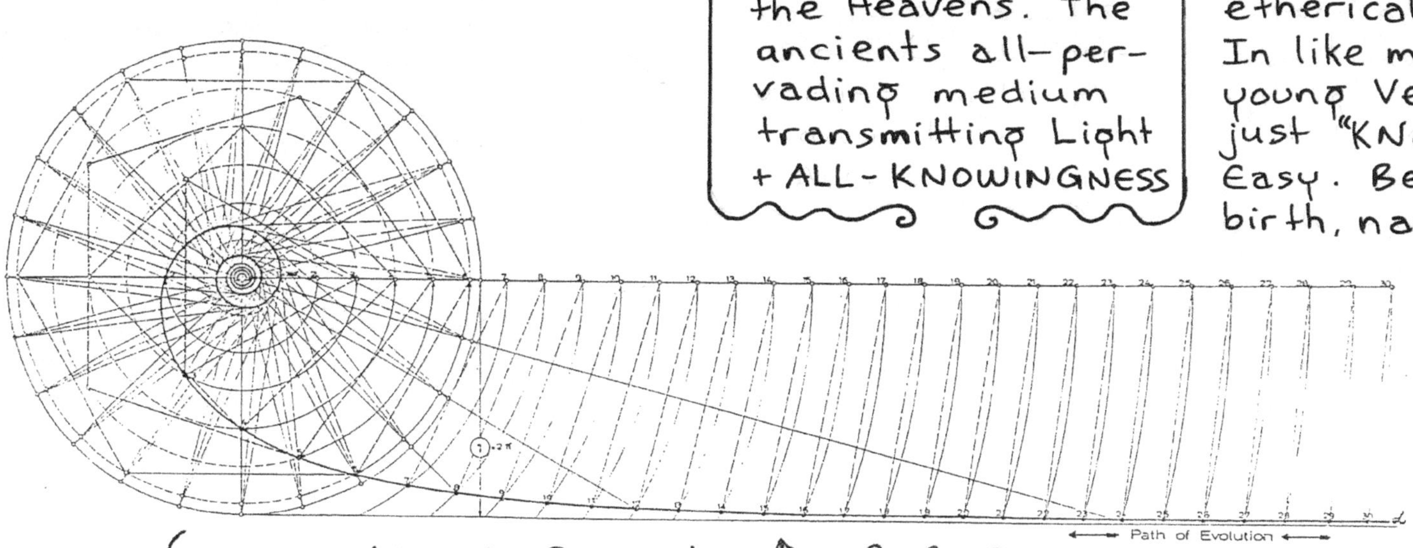

(Inspirational Geometry, by Prof. Callum Coats, to understand the Etheric Blueprint for a damaGed Leaf reGrowinG back to Wholeness.

THE STRUCTURE OF THE MACROCOSMOS AND MICROCOSMOS

THE 'INVOLUTIONARY' SYSTEM — CENTRIPETAL TECHNOLOGY

PYTHAGORAS - KEPLER - SCHAUBERGER SCHOOL

PKS PYTHAGORAS KEPLER SYSTEM

DIGITAL SUMS TO SOLVE 2 DIGITS X 1 DIGIT

Digital Compression is also known as 'Digital Sums' or literally 'Adding Digits' or 'Reduction of Digits'. E.g. any Double Digits like 34, can be reduced to a single digit, by merely adding their sum e.g. 34 = 3 + 4 = 7. Somehow, almost holographically in the sense that each part is a memory of the whole... this single digit '7' holds the vibrational memory of the larger number '34', in a manner like the outer valence electron of a nuclear substance knows the whole of the atom and how it will link or bind with other atoms. Some scientists have dismissed this valuable tool of Digital Compression by calling it 'mere numerology', but in fact its purpose is akin to an X-Ray... penetrating density or randomness to locate where patterning or skeletal structure truly lies. Digital Compression (what I call: the 17th Sutra) allows the invisible to surface into the visible world. This is not just Vedic Mathematics; in fact it goes further back. It is what I call Galactic Mathematics!

Digital Sums to solve 2Digits x 1Digit

Perhaps the most practical form of using this Sutra is in the realm of Multiplication, where any Two Digit Numbers are multiplied by any single Digit.
Eg: 44 x 6, 23 x 7, 24 x 6, 25 x 9 etc
The definition of Vedic Mathematics is 'Mental, One-Line Arithmetic'. I encourage all children to learn this amazingly swift mental method... moving across the line, rather than calculating traditionally down the page. It is of great value to do such sums upon the visualized 'blackboard in your mind'. The ancient Indian civilization practiced mental gazing techniques, called 'Trataka', where the fixed gaze ultimately awakened the Third Eye and gave insight into the science of Complementary Colours in the Artist's Colour Wheel, e.g. staring at a red star for 3 minutes, then closing the eyes produces an after effect of a green star! Thus, 'holding' these numbers in your Mind's Eye, helps build the Mental Muscle. Vedic Mathematics is really about re-learning the lost art of 'Holding Digits on the inner Mental Screen'.

How to mentally multiply 24 x 6 in a few seconds:
Multiply by 6 both digits of 24. Thus upon your mental screen visualise 2 x 6 / 4 x 6 which becomes: 12 / 24.
(note: The forward slash / is used as a place marker only).
Your focus now is to have only one digit on the right-hand-side of the slash.
This Digital Compression is achieved by adding the "2" of the "24" over to the left-hand-side, appearing like this:
12 + 2 / 4 = 14 / 4
When done quickly in the Mind, the mental steps are:
24 x 6 = 2 x 6 / 4 x 6 = 12 / 24 = 12+2 / 4 = 1 4 / 4 = **144**

It's that simple, keep practising with other examples:
4 44 x 6 = 4x6 / 4x6 = 24 / 24 = 24+2 / 4 = 2 6 / 4 = **264**
4 18 x 8 = 1x8 /8x8 = 8 / 64 = 8+6 / 4 = 14 / 4 =**144**
4 71 x 7 = 49 / 7 = **4 9 7**
4 37 x 6 = 18 / 42 = 18+4 / 2 = 22/2 = **222**

This is the **Path of Least Resistance**. You will come to understand that Western Math's is not wrong… just clumsy. Eastern Math's is akin to the manner in which the firing synaptic neuron connections of the brain do indeed prefer to travel linearly along the mental plateau, rather than down the page. In fact research now tends to extend this idea of the linear mental plateau as a holographic image. In fact, we already know the answer. When in the movie called the 'Rain Man' the central character or Idiot Savante was asked:
"How do you perform correctly such complex mathematical tasks"? He replied: **"I see it"**.
This is the **supreme Art of Pattern Recognition** symbolized by the Right Visual Cortex of the Feminine Intuitive side of the Brain. Enjoy this ancient code.

Jain 2004 from the Book: The Vedic Mathematics Curriculum For The Global School.

Jain Products Order Form

NEW! **The Living Mathematics of Nature, Series – 5 DVD Set, pub: 2006**

	Item	Price	Weight
☐	**Complete 5 DVD Set** (Special! Save $25)	**$200**	240 gm
☐	**Introduction to Ancient Knowledge** (DVD #1 of 5)	**$ 45**	100 gm
☐	**Vedic Mathematics** (DVD #2 of 5)	**$ 45**	100 gm
☐	**Magic Squares** (DVD #3 of 5)	**$ 45**	100 gm
☐	**The Divine Phi Proportion** (DVD #4 of 5)	**$ 45**	100 gm
☐	**3-Dimensional Geometry** (DVD #5 Of 5)	**$ 45**	100 gm

Magic Squares

	Item	Price	Weight
☐	**The Book of Magic Squares, Volume 1** (M. Sqs. of 3, 4, 5, 6, 7)	**$ 45**	520 gm
☐	**The Book of Magic Squares, Volume 2** (M. Sqs. of 8, 9, 10)	**$ 45**	520 gm
☐	**The Book of Magic Squares, Volume 3** (M. Sqs. of 11, 12, 16)	**$ 45**	560 gm
☐	**Join-The-Dots** (adults and children coloring book. Pub:1999, 56 pp)	**$ 30**	160 gm
☐	**Decal: The Magic Square of 4 X 4 (Jupiter)** (Small, Magenta)	**$ 10**	20 gm
☐	**Decal: Magic Square 8 X 8 (Mercury)** (Small, Blue Background)	**$ 10**	20 gm
☐	**Decal: Magic Square 4 X 4 (Jupiter)** (Large, Yellow Background)	**$ 20**	100 gm
☐	**The Gospel of the Holy Twelve** (Essene Bible w. Magic Sqs. 210pp)	**$ 45**	410 gm

Vedic Mathematics

	Item	Price	Weight
☐	**The Magic of Nine in Vedic Mathematics.** (Pub:1999, 124pp)	**$ 45**	390 gm
☐	**The Vedic Mathematics Curriculum for the Global School. Part 1, Digital Sums.** (Pub:2001, 142 pp)	**$ 45**	460 gm
☐	**Jain Mathemagics Curriculum for the Global School. Part 2. Multiplication.** (pub: 2005, 152pp)	**$ 45**	480 gm
☐	**Vedic Mathematics for the New Millennium: DVD** (Pub:2002)	**$ 45**	100 gm
☐	**Dolphin Human Connections:** Jain shows Vedic Maths (2006). 40 mins	**$ 50**	100 gm
☐	**Vedic Mathematics** by: Bharati Krsna Tirthaji **Soft Cov.** (355pp)	**$ 45**	460 gm
☐	**Vedic Mathematics** by: ibid. **Hard Cover.** (Pub:1965, 355pp)	**$ 55**	570 gm
☐	**Calculations @ The Speed Of Light** by Kranti Kiran (pub:2002, 104pp)	**$ 45**	230 gm

Sacred Geometry

	Item	Price	Weight
☐	**The Book of Phi, Vol 1** (The Living Maths of Nature. Pub:2002, 172pp)	**$ 50**	550 gm
☐	**In The Next Dimension** (The Book of Phi, Vol 2. Pub:2003, 188pp)	**$ 50**	580 gm
☐	**Propositions on the Exact Value of Pi.** Reddivari Sarva J. Reddy	**$ 60**	620 gm
☐	**Peter's Projection Map.** (Earth's true cartographic proportions)	**$ 25**	50 gm
☐	**Dvd: Jonathan Quintin's "Visual Symphonies"** Toroidal Animation	**$ 45**	100 gm

EarthHeart Jewellery

	Item	Price	Weight
☐	**Platonic Solids:** Crystal Gift Set in wooden camphor boxes	**$100**	50 gm
☐	**Heart Star:** 24 carat gold plated stellated cuboctahedron	**$150**	50 gm
☐	**Tantric Star:** 24 carat gold plated Star tetrahedron	**$130**	50 gm
☐	**Solar Star:** 24 carat gold plated stellated dodecahedron	**$150**	50 gm

SUB-TOTAL: $ ....

Postage & Handling: $ ....

TOTAL: $ ....

Tax Invoice

Name: ....

Mailing address: ....

City/Town: .... **State:** .... **Zip Code:** ....

E-mail Address: ....

Home phone: ....**Mobile:** ....

Please choose your preferred method of payment:

☐ **Cash, Cheque or Money Order.** Amount: ....
Payable: JAIN F.R.E.E.D.O.M.S.

☐ **PayPal now Available!** Amount: ....
Email: jain@jainmathemagics.com

☐ **Direct Deposit** Amount: ....
Acc Name: JAIN F.R.E.E.D.O.M.S. BSB: 032573 Acc Number: 171268

☐ **(circle) Visa / Mastercard** Amount: ....

Card number..

Name on Card: .... **Exp. date:** ....

AUSTRALIAN POSTAGE: Orders up to $50, add $8. Orders $50 to $120, add $12.

INTERNATIONAL ORDERS: $10 set up fee and add $10 per item. (Approx.)
International Bank Draft in Australian Dollars. PayPal also available + Western Union.
Cheques made out to JAIN F.R.E.E.D.O.M.S.

All Prices are listed in Australian Dollars. **All products available online!**

Date Of Order .... **Dispatched** ....

6	1	8
7	5	3
2	9	4

JAIN F.R.E.E.D.O.M.S. inc 9882763 trading as **JAIN MATHEMAGICS**
For **R**esearch **E**xpressing **E**ssential **D**ata **O**f **M**agic **S**quares
Phone: +61 02 6684 4409 Mobile: +61 0423 583 886 ABN 47 244 872 696
Email: jain@jainmathemagics.com Web: www.jainmathemagics.com
Address: 777 Left Bank Rd, Mullumbimby Creek, NSW, 2482, Australia

JAIN MATHEMAGICS CURRICULUM for the GLOBAL SCHOOL

formerly known as: The VEDIC MATHEMATICS CURRICULUM for the GLOBAL SCHOOL

Book 2 MULTIPLICATION JAIN 2005

Developing the Inner Mental Screen

JAIN MATHEMAGICS CURRICULUM FOR THE GLOBAL SCHOOL Book 2 MULTIPLICATION 2005

"THE FUTURE IN MATHEMATICS IS DOING IT IN YOUR HEAD" (by Jain: Mathematical Futurist).

- This is an Educational Workbook for Children and a Teacher's Resource Material for those interested in learning and teaching Rapid Mental Calculation. It increases the child's Memory Power and Confidence.
- This long-waited for book of Ancient Mathematical Short-Cuts excavates many hidden truths and vital properties in the playing field of Numbers.
- There is a rare and unique chapter called "HARMONIC STAIRWAY" from Jain's Dictionary of Numbers, for the first time in print. These choice Mathematical Plums are something that the student stumbles upon. They are working away solving calculations mentally, and when they check their answers in Chapter 2, they are instructed, in the section: DID YOU KNOW, to learn more about that particular number. But these are not any particular numbers, they are Anointed Numbers that our forebears held in high esteem.
- Beautifully and richly illustrated that the graphics alone educate. The student learns about other Families of Numbers, other numerical relatives and mathematical cousins that enrich their understanding of what mathematics is really about. It instills that sense of joy and wonderment that Pythagoras and Baudahayana knew.
- There is no error; it is an infallible bulletproof system, based on Unity Consciousness. The 16 "Threads" or Sutras that solve all known mathematical problems express The Law of Economy and The Path Of Least Resistance.

- As Jain Mathemagics becomes more globally acknowledged, it will help end the generational tyranny that has kept such knowledge in the dusty cupboard. This book is an organic pill that will prevent the slowly encroaching borgificiation of Mathematics.
- Since the turn of the bi-Millennium, there has been a global renaissance in the subject of "Sacred Geometry". You could summarize it as a fascination for the Language of Shape. This book invites seekers to truly incorporate SHAPE into their ability to perform mathematics, to apply this Language of Shape to the next octave of learning, not theoretically, but by practical use of using SHAPE to literally perform mental calculations in seconds.
- Often students come out of school unable to recover from deep mathematical wounds. They are mis-diagnosed as "un-intelligent" or "dumb", given drugs like Ritalin, told they are ADHD (which really means: Attuned Directly to Higher Dimensions). In fact, Dyslexic children are geniuses.
- **UNCOVER, RECOVER, DISCOVER**

BACK COVER

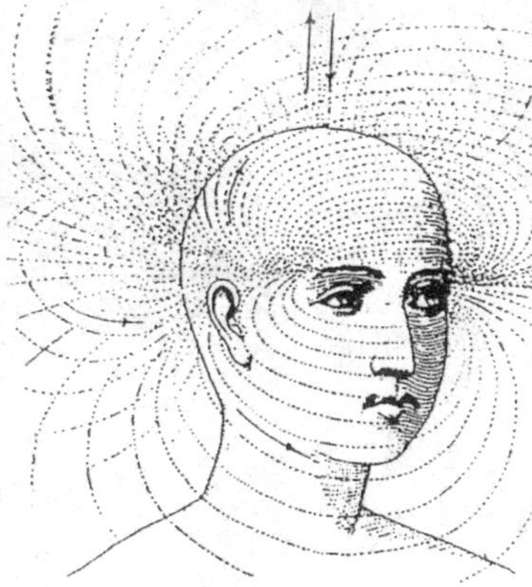

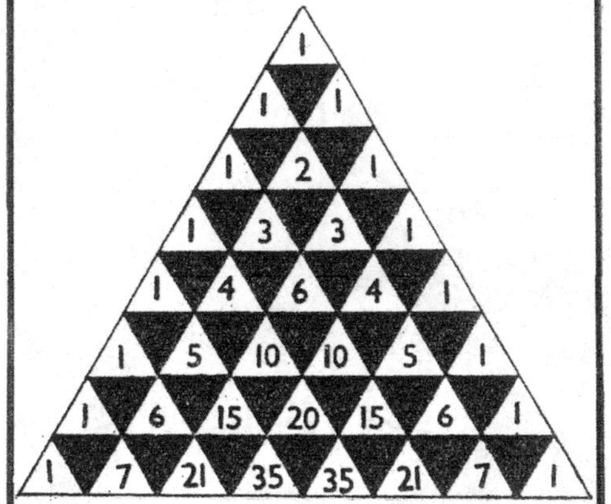

BLURB

- This Book, the first of a 4-part Series, is designed to enhance the present mathematical curriculum, and in some cases, replace tedious or outmoded procedures, bringing joy into this often abhorred subject
- Jain is at the vanguard of advocating mathematical change and reform.

The presentation of these timeless number codes, expressed as "Digit Sums" are part of our genetic memory. It is Right Brain or Feminine Hemisphere Mathematics that sees or intuits the shape or artistic pattern of the predicted answer.

- In this dance of numbers, the dancer will feel interconnected with all branches of knowledge and thus also become a recipient to Higher Knowledge. In this way, the Parent can effectively teach their Child (7 years and onward) the pure first principles of Vedic Mathematics.
- Discover how the Vedic Square, via Digital Sums, reveals Atomic Structures!
- Vedic Mathematics is based on 16 Sutras or word-formulae that were reconstructed by Sri Bharati Krsna Tirthaji, 1884-1960, during an 8 year forest retreat.
- By definition, this universal system is described as Mental, One-Line Arithmetic. It vastly improves memory skills, intuition, mental agility and self-confidence.
- If society continues to remain dependent upon calculators, for the next 20 years, the Pineal Gland of the Global Brain will atrophy. Become a Human Bio-Calculator!

LEARN THE
SECRETS OF NATURE
Mathemagics.com
Jain's 5 Interactive Seminars will help you discover the beauty of numbers in Art, Science, Music and History.
THE ART OF NUMBER Translating Numbers Into Art and Sacred Symbols
RAPID MENTAL CALCULATION Vedic Mathematics Learn how to be a Human Bio-Calculator
MAGIC SQUARES Translating Numbers Into Exquisite Atomic & Mandala Art
THREE-DIMENSIONAL GEOMETRY of the 5 Platonic Solids Reveal Secrets of Atomic Structure
DIVINE PROPORTION Living Mathematics of Nature and the Lost Secrets of the 108 Phi Code
Three Age Groups: Juniors (9-12) Teens Adults
The 5 Day Seminars progress from:
~ Level 1 Introductory
~ Level 2 Advanced
~ Level 3 Teacher Trainer Courses

Made in the USA
Coppell, TX
05 November 2026

11307512R10085